A Transnational Feminist View of Surrogacy Biomarkets in India

Sheela Saravanan

A Transnational Feminist View of Surrogacy Biomarkets in India

Springer

Sheela Saravanan
Department of Anthropology,
South Asia Institute
Heidelberg University
Heidelberg
Germany

ISBN 978-981-10-6868-3 ISBN 978-981-10-6869-0 (eBook)
https://doi.org/10.1007/978-981-10-6869-0

Library of Congress Control Number: 2017955237

Printed on acid-free paper

This Springer imprint is published by Springer Nature
The registered company is Springer Nature Singapore Pte Ltd.
The registered company address is: 152 Beach Road, #21-01/04 Gateway East, Singapore 189721, Singapore

To my parents, my ideals,
R. Suryanarayanan & K.R. Sivakami,
my foremost teachers of social justice.

Acknowledgements

I am deeply grateful to everyone who has helped me in this endeavor. I thank the medical practitioners who allowed me entry into this sensitive zone of commercial surrogacy in India and the personnel in the clinic who helped me meet surrogate mothers and intended parents. I thank the surrogate mothers for their kindness and for sharing their experiences with me and for teaching me how to remain positive and cheerful even in adverse conditions, and the intended parents who shared their experiences with me. Without the financial support of German Research Foundation (DFG), the field work in India and the documentary film would not have been possible. My friends in India, Germany, and Canada who have encouraged me with my work. A few surrogate mothers from my case study were forthcoming in also participating in my initiative on making a documentary film on surrogacy titled "*Mother Anonymous*" directed by Rahul Ranadive. I thank them for trusting me. Their stories, hence, are not only in written format and in photographs, but their voice and presence are also felt through the documentary. I am grateful to inspiration I gained from my feminist friends Renate Klein, Sasan Hawthorne, Janice Raymond and from CADAC—Coordination of Associations for the right to abortion and contraception, CLF—Coordination Lesbienne en France, and CoRP—Collectif pour le Respect de la Personne; Marie Josèphe Devillers, Jocelyne Fildard, Catherine Morin Le Sec'h, Ana-LuanaStoicea-Deram, Francesca Marinaro, and Sylviane Agacinski who brought me to realize the significance of taking a stand; my loving family, my husband V. S. Saravanan and children Sudarshan and Sridharan for their patience. At almost every breakfast and dinner table one of the topics of discussion would be on surrogacy and every chapter, title, and conclusion in this book has been critiqued and commented by them. I thank Amy Clare for editing the first draft version of the book and thanks to all my students and colleagues in South Asia Institute, Heidelberg, with whom I have brainstormed my findings, for their feedback and encouragement.

Contents

About the Author

Sheela Saravanan, Ph.D. has two master's degrees from the universities of Mumbai and Pune, India, in geography and development planning. Her Ph.D. from Queensland University of Technology, Brisbane, Australia, in Public Health was on the influence of biomedical frameworks of knowledge on local birthing practices in India. She has worked and published on the status of reproductive health in South Asia, violence against women, and female infanticide in India earlier and now specializes in new and assisted reproductive technologies (ARTs) in the context of Asia and Europe. Since 2007, she has worked in the universities of Heidelberg, Bonn, and Goettingen in Germany. She has published on global injustice, exploitation, and objectification in the process of commercial surrogacy in India. Since January 2016 she has been working at the South Asia Institute, University of Heidelberg on a DFG (Deutsche Forschungsgemeinschaft) funded project. The research aims to examine individual notions of "desired children" (Wunschkinder/Vansh) shaped by social experiences in the German and Indian contexts that lead to selective abortions. In her ongoing research on prenatal diagnosis, she has applied theories of embodiment, intersectional feminism and ethopolitics in examining the notions of Wunschkinder in Germany and India. She has applied theories of embodiment, global justice, and authoritative knowledge in her previous research. She teaches Global Reproductive Technologies: Socio-Ethical and Legal Dimensions; Theories and Practice of Reproductive Technologies and Feminism and Public Health to undergraduates and graduates studying anthropology at the South Asia Institute, Department of Anthropology, University of Heidelberg, Germany.

List of Photos

Chapter 1
Indian Surrogacy Biomarkets: An Introduction

Abstract By 2009–10, India had become one of the most popular destinations for third-party childbearing due to the lax surrogacy laws, high quality medical facilities, qualified English-speaking doctors and a surplus of women willing to offer their reproductive capacity for money. Stories of surrogacy as bane and boon were rife in the newspapers across the globe. There were success stories of couples with their children on one hand, and cases of exploitation on the other. With several unregistered clinics operating in India, the exact number of clinics or number of babies born were unknown. It was estimated that this was a huge profit-making industry with a business of a €445 million returns. Having lived in India for 25 years of my life, inequalities were not a new phenomenon to me, but nothing had prepared me for the structural inequalities and extreme injustice that I was about to witness as a researcher studying social construction of commercial surrogacy practices in India. The entrepreneurs who flourish in the surrogacy transactions include medical practitioners, agents, hotels, commuting services, landlords of the accommodation, lunch providers for surrogate mothers and many other small entrepreneurs. However, the biggest profiteers among all these are the owners of the fertility clinics. They are profit-making institutions that aim to make surrogate mothers more bio-available than other clinics in the competitive market. They became more competitive on the basis of who could provide the precise service preferred by the intended parents. Eventually, I was drawn into the depths of the lives of the surrogate mothers and the intended parents. I completed my field work in 2010, with a short documentary film 'Anonymous Mother' based on the surrogate mother's narratives and have kept contact with most surrogate mothers.

Keywords Surrogacy farming · Inequalities · Biomarkets · Commodification of children · Hindu epics · Transnational feminism

S. Saravanan, *A Transnational Feminist View of Surrogacy Biomarkets in India*,
https://doi.org/10.1007/978-981-10-6869-0_1

Initially there was almost nothing happening and my field work on surrogacy[1] in India was going at a snail's pace. Sitting endlessly for hours at the waiting area in a small clinic in Gujarat every day and talking to people visiting the clinic was an ethnographical study by itself. The inequalities in which the elite were treated within the clinic as compared to the socio-economically disadvantaged visitors to the clinic was one of my first field observations. I instantly could identify the intended parents[2] with their imperialistic presence and the attention they received from the clinic staff. A group of westerners and elite Indians arrived every day at a particular time in an air-conditioned car; they were taken to a well-kept air-conditioned room attached to a clean western toilet and were soon summoned to the doctor's cabin for their appointment. The poor were sitting outside, in the crowded waiting area, with no air-condition and an unkempt Indian style toilet and were made to wait for hours before they could get an appointment. It was a rule in the clinic that everyone had to leave their shoes at the entrance, but it was amusing to note that some intended parents were reluctant to leave their expensive shoes there and would carry it in their hands into the waiting room.

One woman, sitting beside me held her luggage on her lap. She had come from Madhya Pradesh, after a five-hour train journey, just to meet this doctor on account of her popularity for her infertility treatments. She had an evening train to catch the same day, but despite several hours of waiting and several reminders at the reception she was not called inside. Finally, she had to fight at the reception to be summoned into the doctor's cabin. She was evidently upset when she came out a little later, "she (the doctor) says she cannot treat me, but she requested me to donate my eggs if I want to, why should I donate my eggs, I didn't come here for that." I recollected reading about the unregulated egg donation industry in India, riding on lucrative returns and word-of-mouth. Donors are categorised by looks, height, educational qualifications, with the most significance given to fair skin. Foreign donors, also termed "diva donors", may earn up to INR600,000.

Initially, on my enquiry to conduct a post-doctoral study from Germany, Dr. Nisha willingly gave me permission but when she met me face-to-face she became reluctant probably because she realised that I intended to stay there for a longer period of time to obtain in-depth information about the respondents' lives. Reluctantly, she requested Dr. Harnish, her assistant (also her husband), to help me. He was an orthopaedic doctor himself but as his wife's business was thriving, he had decided to assist her instead. I would arrive early in the morning but had to wait

[1]Surrogacy has also been referred to as gestational carrier contract/agreement, pregnancy contract, which can be demeaning to the surrogate mothers; hence, I have used the term surrogacy, as the women themselves have referred to this practice with these words.

[2]Intended parents refer to the commissioning parents, biological parents, genetic parents, but many intended parents are not genetically related to the child because their involvement in the surrogacy practice begins with their intent to have children in this way; hence I have used this terminology.

until most patients left in expectation to be introduced to surrogate mothers.[3] Most days I would just return without any interview. I was first selectively introduced to surrogate mothers and later met several other women through snowball method. I say selectively because I observed I was introduced only to women who were seemingly submissive, soft-spoken and were full of praises for Dr. Nisha as a demi-god. With my extensive experience in conducting research on sensitive issues in India such as violence against women, working conditions in garment factories and reproductive health, it was not difficult to read between the lines and observe their expressions of admiration. Any doubt that I had was cleared as I proceeded further with my research.

Meanwhile, Dr. Harnish, requested a Dutch intended parent on her way out from his cabin if she would participate in a research. As the door was open I could hear her and she had not identified me because I was sitting along with several others in the waiting area outside his cabin. "Why should I participate, these people write all kinds of things about surrogacy, they say its womb-for-rent", she shouted loudly. Ironically behind her was a clinic notice board on which a newspaper cutting was pinned up; it was an article titled "Wombs for Rent" with a photo of Dr. Nisha, a jubilant western couple, each holding a baby, their twins born through surrogacy with no sign of the surrogate mother on the photo. The other cards and baby photos were the praises, of which I was very familiar by then, "Dr. Nisha, you are our God, we will never forget you all our lives."

One day after all the patients had left, Dr. Harnish called me into his cabin and asked me to wait there. Soon a young girl (Nargisa) seemingly 19 years of age accompanied by a nurse, walked into the room with the intravenous syringe still fixed to her hand. She was a surrogate mother who had just delivered twins two days ago. I wondered how she was here at this clinic and it was then that I found out that one surrogate home was on the second floor of the clinic itself. My second participant was a surrogate mother (Dimpy) who was put up in a children's hospital a few kilometres away from the clinic, along with the baby girl. She and her husband were caring for the baby for almost two weeks waiting for the intended parents to arrive from Turkey. In the clinic, there was a mounting tension about this child as they were unable to contact this Turkish couple for some unknown reason. In my meeting with Dr. Nisha, she told me that the surrogate mothers are well aware that the child does not belong to them and they only are renting their womb temporarily to another couple who would take the child(ren) that rightfully belong to them. She told me that the surrogate mothers only wanted the money and have no interest in the baby and don't (want to) bond with the child. What I saw at the children's hospital was a complete contradiction to what she had told me, the surrogate mother was playing and singing to the baby and even breastfeeding, and

[3]Surrogate mothers are women who carry babies for infertile couples. In India, women are not allowed to use the own gametes, hence are commonly referred to as gestational mothers. Scholars have also referred to surrogate mothers as birth mothers, gestational carrier, contract mothers, surrogates. As women in India involved in surrogacy refer to themselves as surrogate mothers, I would like to use this terminology.

had named the baby too as if it was hers. As I walked out of the children's hospital, one evening, after visiting the surrogate mother, I felt the familiar monsoon winds blowing from the Arabian Sea bringing with it the pre-monsoon showers. The approaching monsoons has always been a joyful time of the year, a relief from the scorching summer heat, but there was nothing familiar or relieving about the *ethnoscape* I was observing in this small town of Gujarat.

Yet another day, Dr. Harnish asked me if I would like to join a baby shower function at one of the surrogate homes. I jumped at the opportunity and joined a group of women already waiting in an auto-rickshaw. This was the second surrogate home managed by this clinic in this small town in Gujarat. It was mandatory in this clinic that all surrogate mothers remain in the surrogate homes from the time the embryo is transferred until the baby is handed over to the intended parents which can last for almost a year. Within the surrogate homes, beds are lined up in a hostel-like environment and women living here have restricted movement lest it hurts the baby. It is more than three decades since Dworkin (1983) wrote about reproductive brothels, wherein technologies similar to animal husbandry would be used on women. She described a farming model within which women will sell their wombs using in vitro fertilization. Women will be held in places similar to prisons where they cannot move freely and will be restricted to a strict standard of behaviour and sell themselves to make babies. This might have then sounded like a tale out of science fiction, but this surrogate home that I had walked into was no fiction. At the surrogate home, it was a big occasion to celebrate the forthcoming baby. A Japanese intended parent and his surrogate mother were sitting in the ritual downstairs. I went upstairs and saw many women lying in beds lined up, they were apparently under bed-rest after the embryo transfer and others were to have restricted movement until the third month of the pregnancy. I couldn't imagine how the women were living in such a house with no possible diversions; there was no television, no radio, no books or computers, no study table, and no newspapers to read. The surrogate mothers ate their food sitting on the floor as the place was filled with beds with no space for a dining table. At the 10th World Conference of Bioethics, Medical Ethics and Health Law, January 2015 in Jerusalem, Dr. Shalev commented, "even prisons have courtyards, but these surrogate homes don't have enough space to walk even within the rooms" (Shalev 2015).

I met many surrogate mothers that day at the surrogate home and they were more than happy for me to come over and meet them again, many gave me their contact numbers. I also met a few intended parents, but only a few agreed to participate in my research. From here on began the ups and downs of the ride, the surrogate mothers seemingly calm but waiting like a dam about to burst to share their experiences with me, the intending parents describing a different narrative, and the medical practitioners expressing a completely contradictory version about the practice. After the first meetings and conducting my in-depth questionnaires, the intended parents found me to be a resourceful person with my multiple language skills. They requested me to help them with local calls, conducting interviews with potential nannies to be taken back to their destination country, and translating conversations between themselves and their surrogate mothers. I maintained contact with them even after they left India with the child(ren). Similarly, I was drawn into the lives of the surrogate mothers and became involved in meeting their children at boarding schools left there because the mother was in surrogate homes, and meeting them after they returned home after relinquishment. In contrast to this small town, in a clinic in Ahmedabad, the surrogate mothers were shifted from a low-income locality to a comparatively higher income locality for the period of the pregnancy, the rent of which was paid by the intended parents. In this new place, they were left with no social support system, especially if they have hidden the surrogacy from their relatives. Here I became involved in supporting the surrogate mother such as caring for her children when she and her husband were at the hospital. The inequalities became even more evident at this stage of my research; I would be eating breakfast in a five-star hotel with an intended parent who would be discussing about the dirty roads and the unsafe food in India and that very night I would be staying in one of the poorest households in the town without basic amenities. It was only when I received a subtle threat because I had unknowingly stumbled into an undercover prostitution racket, also involving some surrogate mothers that I decided to leave this small town. I left Ahmedabad when one surrogate mother (Gayatri) asked me to join a secret mission arranged by an agent to take the surrogate mothers and sell their wombs in Mumbai, but I didn't join this mission as I realized that my data is saturated enough. I have been in touch with the surrogate mothers ever since, however, the intended parents were not happy with my writings and the contact withered away. I realized that two intended parents had agreed to participate in my research with a biased intention to propagate in support of the clinic, which was, by then, already being widely critiqued.

Owing to the sheer complexity of the practice and the data that I had collected, I had been struggling with the idea of this book despite publishing several journal articles. I have shared these narratives by the intended parents, medical practitioners, and surrogate mothers who kindly participated in my study, as well as my experience as a participant observer in Chaps. 5 and 6. Chapter 5 gives a detailed narrative of the surrogate mothers as I met and interacted with them during the course of their surrogacy and post-relinquishment. I conducted my research in two clinics, one in a smaller town in Gujarat with three surrogate homes and one without a surrogate home in Ahmedabad between 2009–10. I interviewed five intended parents, 13 surrogate mothers, and five medical practitioners. Among these, I closely followed five surrogate mothers throughout their pregnancy using participant observation method and hence could interact effectively also with their spouses and family members. I chose to reveal Gujarat and Ahmedabad as my places of study as Ahmedabad is a large city with several IVF clinics; however, I maintained anonymity regarding the small town in Gujarat as it would then be easy to identify the name of the clinic. However, all the names of persons used in the study including the names of the doctors, agents, intended parents and surrogate mothers are pseudonyms to maintain confidentiality.

In brief, my research findings revealed that surrogacy was mainly a money-making business for the medical sector and the agents, who were exploiting not only surrogate mothers but also the intended parents. Surrogacy was a bazaar where everything about women's reproductive capacity and the children born was priced; the woman's body parts, her breast-milk, her labour as a nanny, the number of child (ren) born, the weight of the babies, the gender/(dis)abilities of the child and even the surrogate mother's caste or religion was priced. The first fiction written on surrogacy was Margaret Atwood's *Handmaid's Tale* in 1985, a story about a class of women kept for reproductive purposes by the elite in an era of declining births due to infertility and sexually transmitted diseases. This was followed by Corea and Dworkin drawing parallels between animal and women used for breeding and milking. The surrogacy markets in India was catering to both the western (White and Indian NRI) and the elites within the country who desired this handmaid's services. Corea (1985) and Dworkin (1983) cautioned that commercial surrogacy would move to the poorer countries in the third world or to poverty stricken parts of the USA as women in these places would be willing to do this for less cost. In India, unable to get a decent job, women were driven to sell the only thing that seems to have a value (their body) similar to prostitution. Children too were commodified; they were priced per child, one child had a standard price and every other child (in case of twins/triplets) were charged double the price by the clinic; there was nothing paid in case of miscarriage, disabled/children of undesired sex were left in orphanages, sold or left on the streets in India, as women were paid according to the size of the baby, thereby, women were overfed during the pregnancy in order to increase the weight of the baby.

The terminology used by various participants involved in surrogacy clearly indicated towards a contextual understanding of the practice as a market. The terminology used also clarified the relationship between different participants in the practice. The medical practitioners called the intended parents "clients". This was not by chance but because they earned INR100,000 (approximately 13,000) for one child which is doubled with another child. This may seem cheap when compared to a surrogacy cost in the USA or UK but for an Indian doctor this is a huge remuneration and profit-making option. The surrogate mothers called the intended parents "party *wale*" or buyers. This is again evident in the surrogacy practice as the intended parents should like the surrogate mother in their first face-to-face meeting and it is only then can she be selected for the surrogacy. The surrogate mothers have to abide by whatever the intended parents ask them to do apart from the rules stipulated by the clinic. The surrogate mothers are called as "surrogate" by both the medical practitioners and intended parents, but they call themselves surrogate mothers. They never even by mistake say the word "surrogate" without adding "mother" to it. While for all other participants the mother aspect is anonymized, the money factor is exemplified by the intended parents and the medical practitioners. However, the money factor did play a significant role in this market along with emotional cost (Hochschild 2017). The surrogate mothers earned approximately 3,500–6,500 Euros approximately or INR250,000–500,000, an amount they would take about five to seven years to earn with their present earnings. Although the surrogate mothers were paid a lesser contribution compared to their counterparts in other parts of the world, and the women knew they were exploited, they willingly consented to this. Their financial need and the distress caused by the surrogacy practice was strongly expressed by surrogate mothers in my study. The socio-economic status and basic capabilities of surrogate mothers played an important role in their choice to participate in surrogacy. None of the surrogate mothers in my research had studied beyond higher secondary level and their earning capacity was inadequate. This effected their negotiation powers leading to their involvement in unjust surrogacy arrangements.

Surrogate mothers were mostly women who were already involved in the biomarket, engaged in activities like drug trials and gamete donation, and were easy recruits into the process of surrogacy in India. Realizing that women's reproductive body parts are more remunerative and lucrative an option for quick money-making, they used this opportunity to prove their dedication towards their family. Indeed, most women in my study had willingly consented to surrogacy. Surrogacy agents combed through drug trial centres and low-income communities to recruit women in dire circumstances into this practice. Infertile poor women are completely excluded from the biomarkets (both as consumers and service providers) except for some who were coerced into egg donation. Surrogate mothers wanted to enhance their standard of living and medical practitioners were in this business mainly for profit making and other intermittent agents for earning money.

The intended parents are the other main consumers of commercial surrogacy in India and Chap. 6 of the book narrates their experiences. The clinic charged 20,000 Euros for one surrogacy, which is doubled for twins. Hence, buying capacity was an important factor in this biomarket. The chapter describes the purpose and motives of the intended parents in moving to India, how they chose the particular clinic, their experience with the doctors and the surrogate mothers, and their overall surrogacy experience. This includes the narrative of one intended mother from Canada, another couple of Sri Lankan origin settled in Canada, one German intended father who was stranded in India for 2 years, one German lady living underground in this small town along with her twins, and one NRI couple from Gujarat settled in the USA. I also met other NRIs and intended parents of western origin, who were introduced to me by the surrogate mothers and I have mentioned those experiences wherever relevant. The book reviews the conceptual relevance of neo-colonialism and post-colonialism and the methodological relevance of stratified reproduction and intersectionality towards a transnational understanding of reproductive justice, which leaves some questions unanswered. The book finally explains the significance of "humanitarian assisted conception" by posing the question: "are feminist humane?"

Feminists have been caught up between the universalism versus relativism debates. Postcolonial feminists have criticized the universalization of women globally, the description of the third world women as the 'Other' and for applying western standards of emancipation on all women. The aim of this criticism was to identify the colonial voices within the Feminist discourse that was predominantly Western and for a better understanding of cultural contexts. Subsequently, a postcolonial voice emerged diversifying research focus into cultural determinism along with the existing discourse on universalization. Although contextual studies have effectively aided better comprehension of the ground realities, its isolation from macro-level phenomenon in localized cultural determinism; have also led to the further polarization and reinforcing of Otherness. According to Mohanty (2013), capitalism has emerged as a more urgent concern, hence, she stresses on the need for an analytic framework that is attentive to both the micro-politics of everyday life and the macro-politics of global economic and political processes. The experience of intended parents revealed their control over the surrogacy practice as buyers in a contract and their imperialistic approach towards the life and body of the surrogate mothers whose bodies were assumed to be dangerous and unclean, hence meant to be controlled and needed to be hygenized. Some intended parents emphasized the altruistic motive of the surrogate mothers, considering them as "gifts of sisters", although the hierarchy was clear; one sister was to be the saviour to the other. Other intended parents were clear that the child was theirs, with genetics provided or bought by them, and transferred to the surrogate mother's womb; hence, she had no rights over the child from the very beginning following the geneticization approach. This capitalization and the control over human reproductive biomaterial[4] by the rich using global inequalities and vulnerabilities is a form of recolonization of women's bodies and labour.

In order to set the background of the surrogacy practice in India, I reviewed the markets around the globe and within India in Chaps. 2 and 4. Chapter 2 is a review of the transnational movement (global North–South as well as within North–North and South–South) of biomaterial, people, and equipment providing assisted reproductive technologies worldwide. The chapter reviews this movement through cases and the stream of events that occurred in various countries resulting in the prohibition of surrogacy in some Asian and African countries. One important ethical concern of this transnational movement is the development of biomarkets, in which certain bodies become more bioavailable within the existing global and structural inequalities. Such global inequalities are evident not only in the transnational movement for surrogacy but also in similar biomarkets such as gamete donation, organ donation, trafficking and prostitution. These markets raise ethical questions of exploiting the needs of the poor particularly where disadvantaged participants enter into unjust contracts, its relevance to informed consent, unequal distribution of health resource, unfair distribution of benefits, violation of good medical practices, and commodification of women and children.

Surrogacy began in a big way in a small town in India when a elderly lady became a surrogate mother for her daughter and son-in-law. This was a case of altruistic surrogacy at Anand, Gujarat; where a lady of Gujarati origin and living in the UK, experiencing infertility, impregnated her mother using her genetics. Although the case received acclaim, the idea about this surrogacy came from the doctor of the clinic who asked the daughter to request her mother to become a surrogate. At first the daughter was shocked but eventually she consented to the idea. The mother too was apprehensive and on persistent convincing by the doctor and the daughter she consented. According to the doctor at Anand, "she (the

[4]Human biomaterial refers to all biological material such as oocyte, sperms, stem cell, tissues, breast milk and the surrogate mother's womb.

grandmother) was very apprehensive about other people's reactions. But I prevailed on her. In the end, she agreed because she wanted to make her daughter happy. Now she is very happy with what she has done" (Bhatia and Oakeshott 2004). The surrogate mother, in this case, was the grandmother, a live-alone elderly woman in Gujarat, her husband had another home in the UK, depended on her children for her livelihood. Can this case be considered absolutely altruistic without any coercion or dependency? After this case was popularised by the media, India being a country with an estimated 260 million people living below the poverty line, surrogacy caught momentum as a commercial practice with women as an easy prey and thus began a vicious circle of suggestion, coercion and consent.

The surrogate homes themselves emerged from some of the fears of the "Other" expressed by the intended parents as well as the medical practitioners, few being mentioned here—would the surrogate mother get infected with a communicable disease if she continues with a physical relationship with her partners and then transfer this infection to the child; would her (assumed unhygienic) living condition affect the health of the child; would the child's palate get effected by her eating habits (spicy food); if she lives in dirty surroundings or eats unhealthy (spicy) food would the baby growing in the womb be affected; if she has physical relationship with her husband post-embryo transfer would there be chances of a mix-up in the genetic linkage of the child; and would they blackmail the intended parents with the pregnancy. These were the primary concerns besides, some surrogate mothers wanted to stay away from their house in fear of stigmatization. But it is mandatory for the surrogate mothers to stay in these homes based on the demands of intended parents. Being confined to surrogate homes during their pregnancy thus meant women are denied participation in public life and cannot meet their non-reproductive aspirations be it educational or occupational and have a normal social life. The intended parents along with the advice from the medical practitioners created and controlled fixed deposits in the name of the surrogate mothers. All monetary transfers were made strictly through the clinic and they decided the justified reasons where the money should be spent. This Kantian philosophy of utilitarianism within the surrogacy practice normalizes violations of basic human rights wherein the surrogate mother becomes treated as a means to an end. Some scholars highlight the advantages of living in such homes, facilitating networking, bonding, and togetherness (Pande 2014). Whether the mandatory stay in surrogate homes is an advantage or a violation of human rights is discussed further in Chap. 5.

Elly Teman says that surrogate mothers may be submitting their bodies to medical control but they actually use this technology for their own benefit which would be discussed further in Chap. 5. Teman (2010) observes that surrogate mothers may believe in the essentialism of nature but also exemplify the empowered idea independent of the influence of nature (by alienating herself from any feelings of motherhood) also described by Thompson as "agency through objectification" (Thompson 2005: 179). These scholars see the agency of women allowing the power of technology to control their bodies through objectification. Women attempt to prove that they are in control over any innate emotions this procedure may have on the assumed natural emotions (of motherhood) that may stem from their bodies. By submitting to patriarchy, medicalization, technology, and objectification, surrogate mothers "make sure that 'maternal nature' manifests itself only where and when they want it to" (Teman 2010: 53). According to this argument, women should be able, and allowed to find the limited agency by submitting to the justified control of patriarchy and medicalization and in the process, should hand the control of their body over to technology in order to prove that they are not submitting to nature (maternal emotional) and can have control over their feelings of motherhood. Whether this characterizes liberation or reiterates alienation, inequalities and objectification will be discussed further in Chap. 5. Too much focus on unprogressive subjectivity and micro-level autonomies that women exercise within the process of surrogacy mystify the larger picture of structural inequalities and injustice.

The flow of people reveals a pattern mainly from the global North to the global South with some intended parents from within India. It is estimated that 60–90% of the intended parents who commissioned surrogacy in India were from abroad (Bhalla and Thapiyal 2013; NDTV 2015). I was told by the surrogate mothers that NRIs continue to find a way to commission surrogacy in India. The business is very much inclined towards catering to the global North. Recently, Julie Bindel, a journalist from the UK approached a clinic in India along with her friend posed as her sister-in-law of Indian origin requesting possibility of commissioning surrogacy and the clinic was willing to oblige (Bindel 2016). "As soon as we come to a poor country, everyone talks about exploitation" says Dr. Nayana Patel. There is over-reaction when people from global North come to global South for surrogacy, states Pande (2014). The concern is that this outsourcing is being normalized, necessitated and naturalized in the name of liberation of women. I will discuss this in more detail in Chap. 7 and question whether this technology is liberating or not from a feminist perspective. But different forms of normalization of surrogacy is manifest especially among the upper-middle class and elite class around the globe.

Surrogacy is normalized by likening it with any other form of labour and reducing it solely to altruism, "gifts for global sisters", and being considered as a legitimate way of having a baby. While academics, feminists, ethicists, and disability rights activists have cautioned and critiqued it as a social practice that embodies and reinforces social prejudices and forms of discrimination such as imperialism, classism, racism, sexism, and ableism (Peña-Guzmán and Crozier 2016). It is capitalism, flourishing in inequalities that drives the global surrogacy markets, whereas the markets in countries dependent on altruistic surrogacy are almost negligible with hardly any contenders. In India, the cinema fraternity using surrogacy, despite having biological children, also play a role in normalizing the practice through public media inspiring the elite to follow suit.

Many scholars try to normalize surrogacy also in terms of religion by referring to examples from the Bible and Mahabharata (Katz 1986; Rozée et al. 2016). In a TV debate on surrogacy Dr. Nayna Patel referred to Hindu religious texts and said, "go to the Hindu mythology where there are incidences of surrogacy, well accepted, Balarama was born to a surrogate mother so it is a well-accepted fact that surrogacy did exist ages ago" (NDTV 2015) However, according to the texts, Rohini (the surrogate mother) of Balarama is generally considered and respected as his birth mother. It was Rohini who raised him and played a significant role in his upbringing. This is not comparable to the present-day market situation where babies born to surrogate mothers are separated soon after birth and surrogate mothers are supposed to sign off all rights over the child(ren) and all contacts between the intended parents and the surrogate mothers is severed in return for money.

Another form of normalizing is by focusing on altruism and sisterhood in the surrogacy practice. Surrogacy is necessitated by classifying infertility as a form of disability, a social burden that demands a solution at any cost. It is a socially stigmatised phenomenon that reinstates pronatalism and geneticises parenthood. Surrogacy is naturalized through geneticization, thereby prioritizing genetic links and devaluing the gestational role. The rhetoric used to describe surrogacy include "wombs in labour" wombs as a "vacuum", "containers", "rented wombs", a "waste" of reproductive capacity that comes to the use of others. In the documentary film *Made in India*, Dr. Mrs. Kaushul Kadam, Rotunda Clinic said, "I educate these surrogates, I'm just going to prepare a baby outside and put it into her uterus. I only need her uterus, that's when they are able to understand" (Haimowitz and Sinha 2010). First, this indicates that the surrogate mothers were generally unaware of the procedure and second, they are considered merely a vessel for carrying the baby. Separating the role of gestation and childbirth from the remaining body is a form of alienation. Anyone experiencing or who has seen others experiencing childbirth knows that parturition, labour pain, and childbirth if not psychologically, physically impacts each and every nerve of the body and not just the womb.

While the intended parents prefer India because of lower cost, the more important reason they come to India is because the surrogate mother has lesser rights over the surrogacy process and over the child and there are several women who are bioavailable as surrogate mothers due to poverty. A surrogate mother in the documentary film, *Made in India* said, "It's only because of poverty, otherwise I would never have taken such a big step". India ranks approximately 135 on worldwide human development indicators (HDIs), other countries commonly providing surrogacy services also have relatively lower HDI ranks (e.g. Nepal 145, Ukraine 83, Republic of Georgia 79, and Mexico 71) (UNDP 2014). There are also countries in transitional economies with high income inequalities. Reference to poverty is important because most of the women serving as surrogate mothers in India do so to provide their family with immediate basic human needs and adequate food. While some are living on the edge of poverty wanting to provide for their children's education, to buy a house and to avoid slipping further into poverty. Dworkin (1983) notes that it is "the state (that) has constructed the social, economic, and political situation in which the sale of some sexual (prostitution) or reproductive (surrogacy) capacity (becomes) necessary for people's survival". This situation denies people of a host of other possibilities, from education to jobs to equal rights before the law. Hence, there should be more focus on providing women with all these basic entitlements and human rights so that she doesn't have to sell her body in the first place (Dworkin 1983). Protest needs to be directed towards enhancing the essential needs of people in transitional economies like India and not towards encouraging women into surrogacy.

The feminist movement has made commendable progress in enhancing women's reproductive rights and freedom of choice making use of medical technologies such as contraceptives, caesarean sections, and abortions. However, in the last two decades, developments in reproductive technologies and its use for practices such as sex selective abortions and surrogacy has challenged the very ideologies that feminism fundamentally represents equality (socio-economic, health, legal), liberty (freedom of choice, autonomy), and justice (social and reproductive). Liberal feminists support the idea of procreative liberty and individual rights to enter into any contract that one may desire. On these grounds, harm reduction and harmonization has been presented as a solution to reduce infertility on one hand and inequalities on the other.

I argue from a context-specific, power-reflexive, postcolonial perspective drawing on stratified reproduction as an analytical framework that there is a violation of human rights. Using another individual's multiple oppression to fulfil the baby desires of the elite couples cannot be a reproductive right or liberty. Those seeking surrogacy arrangements face social stigma, psychological problems, physical stress of infertility treatment, and violation of bodily integrity. However, opting for surrogacy is likely to put another woman (the surrogate mother) through the same set of problems; social stigma, psychological challenges, violation of her bodily integrity, and also, put the surrogate mother's health, freedom, liberty, and even life at stake. Violation of another person's dignity, integrity, economic needs, hence, cannot be a constitutional right. Economic motivations, a contract out of dire monetary needs, cannot be defined as free choice, instead, more attention should be paid to providing families with the basic human needs such as food, health, education, and housing.

Technologies such as surrogacy provide a wider reproductive choice only for affluent people at the cost of health, freedom, and life of some others (mostly the less affluent) while designating substantial control and power in the hands of intermediate agencies in control of the technology. The socio-economically disadvantaged people in society regardless of whether they are married, single, infertile, same-sex couples, or eunuchs cannot have children through surrogacy whether it is in India or any other country in the world. Any form of individual liberty that seriously impinges another's health and freedom does not conform to the reproductive justice framework. Altruistic surrogacy has been considered a better approach to reduce commercialization; however, altruistic contracts also have elements of structural inequalities and previous experience from other countries, like the UK, reveals that considerable amounts are transferred in the name of medical bills in this process.

A reduced focus on broader global and structural inequalities and elements of racial, sexist, classist, and socio-historic oppressions has led to a felt need for a transnational feminist alliance. Transnational Feminism as a concept is not new and several scholars have suggested it as a way forward to build a global feminist solidarity. However, specific frameworks towards building such an alliance have not emerged within this discourse. I suggest that an alliance drawing on the reproductive justice framework developed by ACRJ (2005) along with the Sister Song Collective has the potential to recognize and address inequalities and injustice observed in surrogacy practices. Reproductive justice is a human rights and social justice based framework that links women's reproductive and sexual health and their social, cultural, economic and political conditions, taking into account the intersecting socio-economic factors, race, religion, geography, nationality, sexuality, and overall health aspects that affect women's lives. This framework aims at recognizing the histories of reproductive oppression in all communities, identifying and addressing multiple oppressions of race, class, gender, sexuality, ability, age, and immigration status to bring about changes in the structural power inequalities and develop political solidarity between women across class, race, ethnicity, sexuality, and national border. Reproductive justice thus aims to reduce inequalities and not to use someone's vulnerabilities as a solution for infertility and thereby create a pathway towards determining global human rights and eliminating discrimination against women. Some of the processes and practices of assisted reproductive technology seems to be crossing the boundaries of humanitarianism and also the very ideologies that feminism itself stands for. Chapters 7 and 8 works towards identifying the humanitarian thresholds of ART practices such as, PGD for sex selection and surrogacy and its global relevance to human rights and feminism. Feminists are being described as inhuman in recent times and some of the feminist discourses reflect these insensitivities towards fellow human beings in general and women in specific. This chapter begins with holding individual reproductive rights and liberties responsible towards a just and humane society by identifying the humanitarian thresholds of feminism and examining where the feminism discourse is crossing such thresholds.

References

ACRJ. 2005. A new vision for advancing our movement for reproductive health, reproductive rights and reproductive justice. Asian Communities for Reproductive Justice. http://strongfamiliesmovement.org/assets/docs/ACRJ-A-New-Vision.pdf. Accessed 14 Aug 2017.

Atwood, Margaret. 1985. *The Handmaid's tale*. Toronto: McClelland & Stewart.

Bindel, Julie. 2016. Julie Bindel speaking at feminist conference against surrogacy. *Byline*. https://www.byline.com/project/43/article/820. Accessed 15 Aug 2017.

Bhalla, Nita, and Mansi Thapiyal. 2013. India seeks to regulate its booming 'rent-a-womb' industry. *Reuters*. http://www.reuters.com/article/us-india-surrogates-idUSBRE98T07F20130930. Accessed 14 Aug 2017.

Bhatia, Shekhar, and Isabel Oakeshott. 2004. *Daily Mail Online*, 2nd. http://www.dailymail.co.uk/health/article-206966/Woman-gives-birth-grandchildren.html. Accessed 15 Aug 2017.

Corea, Gena. 1985. *The mother machine: Reproductive technologies from artificial insemination to artificial wombs*. New York: Harper and Row.

Dworkin, Andrea. 1983. *Right-wing women*. New York: Perigee Books.

Haimowitz, Rebecca and Vaishali Sinha. 2010. *Made in India*. DVD.

Hochschild, Arlie. 2017. Money and emotion: Win-win bargains, win-lose contexts, and the emotional labor of commercial surrogates. In *Money talks: Explaining how money really works*, ed. Nina Bandelj, Frederick F. Wherry, and Viviana A. Zelizer, 161–170. Princeton: Princeton University Press.

Katz, Avi. 1986. Contract motherhood and the baby-selling laws. *Columbia Journal of Law and Social Problems* 20: 1–53.

Mohanty, C.T. 2013. Transnational feminist crossings: On neoliberalism and radical critique. *Signs* 38: 967–991.

NDTV. 2015. We the people—The surrogacy debate; of motherhood or money? *New Delhi Television*. https://www.youtube.com/watch?v=7OlBHVKZzjg. Accessed 14 Aug 2017.

Pande, A. 2014. *Wombs in labor: Transnational commercial surrogacy in India*. New York: Columbia University Press.

Peña-Guzmán, D.M., and G.K.D. Crozier. 2016. Surrogacy as medical tourism: Addressing global inequalities in surrogacy. In *Handbook of gestational surrogacy: International clinical practice and policy issues*, ed. E. Scott Sills, 46–54. Cambridge: Cambridge University Press.

Rozée, Virginie, Sayeed Unisa, and Elise de La Rochebrochard. 2016. Gestational surrogacy in India. *Population and Society* 537: 1–4.

Shalev, C. 2015. Panel chairperson. 'Ethics and regulation of intercountry medically assisted reproduction'. In *UNESCO Chair in Bioethics 10th World Conference on Bioethics, Medical Ethics and Health Law*, Jerusalem, Israel, 6–8 Jan.

Teman, Elly. 2010. *Birthing a mother: The surrogate body and the pregnant self*. Berkeley: University of California Press.

Thompson, Charis. 2005. *Making parents: The ontological choreography of reproductive technologies*. Berkeley: University of California Press.

UNDP. 2014. *Human development report. Sustaining human progress: Reducing vulnerabilities and building resilience*. New York: United Nations Development Programme.

Part I
Surrogacy Globalscape and the Feminist Discourse

Chapter 2
Surrogacy Globalscape

Abstract The liberal feminist approach emphasizes that reproductive freedom is rooted in individual choice and autonomy. Assisted reproductive technologies (ARTs) are perceived as an ideal measure to solve infertility. Importance is given to individualistic anti-statist approach, aiming to provide free access to all kinds of technologies to all individuals. Such open access is meant to be an important component of women's empowerment. What is overlooked in this approach is that autonomy may be provided at the cost of marginalized people's health and well-being. The liberal approach limits embodiment primarily to an individual level, hence, it is inadequate in analysing the complete social complexity of ARTs. Individual decisions are made within socio-political and economic contexts and people's experiences are embedded in communities and histories. Although individual analysis is important in understanding micro-level complexities, it is also important to place individuals and communities within a macro context to be able to understand the broader patterns of marginalization and empowerment. This chapter examines the broader patterns of the surrogacy markets, its access and movements globally with an aim to understand its macro-level global reproscapes. Inhorn (2011) noted the relevance of Appadurai's theory of global scapes to understand the cross-border landscape of assisted reproductive technologies. She reframes it as reproscapes and adds bioscape (moving biological substances and parts) to Appaduarai's five global scapes; people (ethnoscape), technology (technoscape), money (financescape), images (mediascape), and ideas (ideoscape) (Appadurai 1996). Another layer of legalscape (laws and legalities) can also be added to the scapes.

Keywords Surrogacy global reproscapes · Procreative autonomy
Neocolonialism · Postcolonialism · Geneticisation

Liberals consider procreative autonomy of the intended parent's use of technology for 'any' reason that would realize the couple's reproductive goals and enable them to enter into a contract as legitimate (Katz 1986; Robertson 1983, 1986; Andrews 1988; Shalev 1989; Shultz 1990). Other feminists, ethicists and disability rights activists question the 'any' in this reasoning as reinforcing racist, classist, ableist,

S. Saravanan, *A Transnational Feminist View of Surrogacy Biomarkets in India*,
https://doi.org/10.1007/978-981-10-6869-0_2

sexist, nationalist, and casteist prejudices (Field 1988; Pateman 1988; Rothman 1989; Okin 1990; Anderson 1990). This chapter reviews traces of the naturalized, normalized, and necessitated forms of these prejudices in the global surrogacy markets (three Ns). The global pattern of surrogacy markets reveals a significant demand-based flow of people from the global North towards global South along with the flow of desired gametes from the global North; however, there are nodes of flow evident also within Europe and between Europe and America. Before I examine this global pattern, I begin with a review of the broader contexts of naturalization (denoting prioritization of genetic links) that tends to normalize several social prejudices. These prejudices include racism (deems some genetics as superior), alienation (the surrogate mothers from any embodied parenthood), and classist exploitation (considering surrogacy as any other paid labour like a garment factory worker or housemaid).

Geneticization, Racism, and Ableism: Naturalized and Normalized

Racial prejudice is casually applied in the IVF procedure wherein an online donor selection based on desired phenotype is followed by request and subsequent gamete extraction from the specific selected donor. Dr. Richard Sherbahn, who founded the Advanced Fertility Center of Chicago in 1997, says that one couple recently came to him looking for “a thin, white woman with green eyes and wavy blond hair who stands between 5-feet-7 and 5-feet-8, has Swedish ancestry, played college-level athletics, holds a master’s degree at the very least, and registers a minimum 120 on an IQ test” (Hess 2014). In May 2013, a Delhi-based couple in their forties had twins through a surrogate mother, who carried an embryo created from the husband’s sperm and eggs from a Caucasian donor. Through a foreign agency, an unmarried woman was requested to come to India to donate her eggs for which she was paid INR8,00,000. The fertilized embryo was transferred to an Indian surrogate mother who delivered the baby. “The husband is very fair and they wanted to have a fair child. We could not get an egg donor who could match the husband’s complexion and, hence, a Caucasian donor was selected,” said Dr. Duru Shah, who runs a private clinic at Peddar Road. It was revealed that “the donor is a student who wanted to fund her nursing course and so agreed” (Vora 2013). European donors may charge between $1,000 and $5,000 (INR6,000–INR30,000 approximately) depending on factors such as physical health and educational background. Higher caste and fair skinned surrogate mothers have more choices of being recruited in clinics and can claim higher payments (Dhar 2012; Nelson 25 October 2012). Beautiful and fair, the choices used in selecting brides in India are also used in selecting surrogate mothers. Global migration is increasing at a rapid pace, and to an extent that has never been observed before. This racist parenthood pattern is increasingly affecting adoption practices. Recently, an Indian couple living in the

UK was denied their adoption application at a British agency on account of their race; instead, they were advised to try for a brown skinned child in India (Hindustan Times 2017). Using the term "racialized gender", Amrita Banerjee (2014: 113) expressed apprehension in "women's reproductive labour becoming increasingly stratified within the global economy along racial and other lines".

The preference of intended parents for specific genetic material based on phenotype and characteristics of the donor has generated a demand-oriented egg and sperm bank market in Denmark, Switzerland, Czech Republic, Spain, Belgium, the Caucasian countries, and the United States. Spain has become a popular location for Germans seeking pre-implantation genetic diagnosis (PGD), and gamete donations due to their permissive laws passed since the 1980s (Pfeffer 2011). The demand for specific characteristics of donors hence creates a supply channel in which certain donors and their biomaterial become more bioavailable over others. There is a heterogeneous donor pool in Spain as donors are recruited from Eastern Europe in order to suit the needs of Northern European recipients and because Spanish law obliges clinics to match similar phenotypes (Bergmann 2011). This may seem a genuine selection of gamete so that children will look similar and experience a sense of belonging with their parents; the downside of which is that this reinforces racist prejudices, with differing market rates and pricing based on which gametes are considered superior. It is a well-known fact that human eggs of white-skinned women are worth more than those with brown and black skin, hence, the entire baby business is based on racism and colonialism. In recent times, gamete donation seekers are also known to prefer intelligence over looks (Henig 2014). Whether this practice ensuring a demand-based or supply-based market, or is an influence of the rapid advancement of technology or a combination of all three are not the questions I am primarily interested in. I am rather concerned about the reinforcement of prejudices, the domination of some bodies over others, and the systematic subjugation of some bodies that follow the historic patterns of oppression.

Intended parents express a desire for children with certain physical characteristics through their selection of gametes based on the donor's skin colour and education. Surrogate mothers in India are not allowed to use their oocytes. It is assumed that there is no genetic transfer in the gestation phase, hence, the surrogate mother's role is commonly referred to as a "gestational mother". However, medical science has brought to our knowledge that cells migrate during pregnancy and this exchange occurs not only from the mother to the fetus but also from the fetus to the mother (Dawe et al. 2007). So babies and surrogate mothers exchange DNA (Deoxyribonucleic Acid) material during pregnancy. There is a debate among scientists whether or not DNA can be accepted as genetic material and if it is so then all surrogate mothers should be considered "birth mothers" and the entire definition of "gestational surrogacy" and descriptions of wombs as containers need to be questioned. Hence, even from a medical science perspective, the birth mother cannot be completely alienated from the child and should have parental rights over the child. Moreover, erasing a birth mother from her maternal identity and denying her such rights within surrogacy contracts can only be possible under the garb of patriarchy (Cornell 1998). Expectations of certain governments that surrogacy mothers should sign off her rights over the child even before the pregnancy, hence, are completely unjustified.

The biotechnological revolution demands a change in our understanding of embodied race, ability, gender, and nationalist specificity through kinship identities. It is through gamete selection, prenatal diagnosis, and selective abortion practices that culture and ethnicity is inscribed on bodies and perceptions of nationality. Campbell (2007) raises some important questions regarding the new forms of kinship through new reproductive technologies (NRTs)—are the phenotype choices feeding into the larger racist desires through clinical means? Are these deliberate assemblages of bodily matter, identity, and desires also changing the conceptions of national assemblages? Differential regulatory paradigms are defined by social acceptability through perceived and embodied exchange of meanings of risks and symbolic boundaries such as biological citizenship, kinship, and religious perceptions (Rose and Novas 2005; Lamont et al. 2015). Rabinow's (1996) concept of "bio-sociality" explores new forms of identity making on the basis of genes, phenotype, and genotype. Campbell (2007) draws analogies between human fertility regulations, and immigrant policies in the UK, Spain, and Norway, and how the "raciological" (Gilroy 2000) underpinnings of kinship and nationhood is normalized and assumed to be "given" (Gingrich 2004). Gilroy (2000) calls this the "*crisis of raciology*" and asks for a total abandonment of the concepts of race.

Lippman (1991, 1994) conceptualized geneticization as an ongoing process by which individuals are reduced to and differentiated by their genetic identification and capacities. The concept encompasses an expansion of health and illness through genetic technologies; differentiation of individuals based on their genetic variation; construction of biological phenomena through inappropriate labelling of health and disease as "genetic" rather than social, structural or environmental, political economy that uses genetic information to predict and prevent disease; and socio-cultural context that reinforces the use of genetic technologies, especially in terms of women's reproductive choices. Feminists contend that geneticization naturalizes the oppressive use of biology through genetic reductionism which explains human traits in terms of genetic characteristics and genetic essentialism assigning human identity and function as absolute genetic attributes. The new forms of kinship, race, and gender are not merely changing the concepts of nature and genetics but they are also vigorously reinstating the pre-existing beliefs about nature in the forms of categories and distinctions. The desire to have a genetically related child is influenced and constructed by society (Roberts 1996) and it is our responsibility as humans to question and do away with this racist normalization reinforced in assisted reproductive technologies (ARTs). Several feminists have advocated for abandoning the naturalization of genetically linked parenthood as it reiterates patriarchy. Legal custody of children born through surrogacy is usually given to parents with genetic ties, on account of the rationale that they have commissioned or paid for the surrogacy, downplaying the gestational role of the surrogate mother. I suggest at the end of this book possible means to do away with some of such normalizations.

Alpers and Beckwith (1993) expresses concern that price tags in the baby business is based on (assumedly imperial definitions of superior) phenotypes, skin colour, abilities, and characteristics of the donors, which reinforce various forms of prejudices and discrimination. Nelkin and Lindee (1995) argue that "genetic essentialism" reduces individuals to a molecular entity, identifying human beings in all their socio-historical complexities with their genes. When surrogate mothers are selected on the basis of their appearance, caste, religion, and body weight, it introduces new forms of geneticization into the baby business. Genetic selection identifies the gene as being central to human personhood, identity, and social relationships. When most court cases hand over the custody of the children to intended parents based on the genetic determination of parenthood, despite a request of custody by the surrogate mother, it reiterates geneticization. The significance given to geneticization and genetic essentialism, the meanings given to genetic links through the ownership over the gametes and gestation, the gestating body, and the babies born, and such naturalizations of filiality, bring us face-to-face with the memories of nineteenth-century raciological biology that haunts rhizomic theories of hybridity.

"No parent wants a son with a disability" was the statement of the Australian couple, who commissioned surrogacy in India and abandoned one of the twins for having down syndrome and returned to their home with the other twin, a baby girl. Another UK parent abandoned a disabled daughter and kept the son after her surrogate mother gave birth to twins (Baklinski 2014). In my study conducted in Gujarat, one baby girl was abandoned on the streets of Ahmedabad one week after birth by the intended parents; the surrogate mother recognized the baby but these cases go unreported. These incidents are a consequence of broader forms of normalized ableism that sociologists have cautioned about, that entails discrimination on genetic grounds (Dreyfuss and Nelkin 1992; Nelkin and Tancredi 1994; Paul 1994). Some have suggested that surrogacy is a rebirth of scientific racism and eugenics driven by consumerism and reproductive choices that feeds into parental desires of a perfect child (Duster 1990; Rifkin 1998; Hubbard and Wald 1999). The capacity of biomedicine and biotechnology to improve health, welfare, and quality of life is celebrated while the capacity of the same technology to control, coerce, restrict, and even eliminate the so-called defective through individual practitioners and political authority needs to be questioned (Rose 2001). Agamben (1998) explains the biological paradigm of the modern wherein exclusion of defective individuals becomes essential, hence, normalized.

Social and historical contexts also define meanings in which women's bodies are controlled and reproductive bodies turn into mediums of racist values. An Ohio woman sued a sperm bank after it mixed up donors and gave her the sperm from an African–American man instead of the white skinned gamete she and her partner selected for. They had paid and selected a donor with blonde hair and blue eyes so that they could raise a child that bears some resemblance to them. They say they had not planned to raise a mixed-race child as they were living in a conservative and racially intolerant neighbourhood. This may sound absolutely genuine but raises serious questions about the connotations of naturalization and how our bodies are colonized by racist histories and in contrary to the concept of reproductive liberty (Petchesky 1995).

Surrogacy Necessitated as a Solution to Infertility

The market is necessitated, along with being normalized and naturalized. The necessitation of surrogacy is justified by presenting infertility as a disease, a disability. Childlessness is socially constructed as a malady, arising from patriarchal hegemony of social norms pressurizing especially women to use ARTs. This myopic social and medical focus on women's body to solve infertility and the perpetuation of objectification and commodification of women's bodies is to fulfill patriarchal and commercial ends (Raymond 1993; Corea 1985; Harding 1991). When infertility occurs, pronatalist and heteronormative identities and norms result in social stigmatization due to pressures to conform with it. Women in different cultures face a diminished sense of identity and social standing in the community on experiencing infertility (Donchin 1996). Germany is known to be a strong pronatalist country and women (and couples) face stigma if they do not reproduce (Greil 1991; Riessman 2000). Infertility and childlessness are considered unnatural and lead to a perceived sense of "an 'unfulfilled' life for the woman [and] 'emasculation' for the man" (Donchin 1996; Benninghaus 2012). In India, there is a strong preference for heteronormativity, pronatalism, and the notion of compulsory motherhood (Pujari and Unisa 2014). Thus, women (and men) in Indian society, defined by their fertility, can feel worthless when faced with involuntary childlessness and would thus do all they can to conceive a child of their own (Dhar 2013). It is important that these prejudices related to infertility are addressed, while the focus of biotechnology is on providing the best services according to people's desires by offering a range of choices that subtly reiterates these social vices.

Almost every article advocating for surrogacy begin with estimating the number of infertile couples or persons with infertility. A recent estimate of the number of couples facing infertility is that it increased from 42.0 million (39.6 million, 44.8 million) in 1990 to 48.5 million (45.0 million, 52.6 million) in 2010 (Mascarenhas et al. 2012). However, global infertility prevalence rates are difficult to determine, due to the lack of consistent use of definitions and common tools to diagnose, manage, or report infertile individuals and couples worldwide (WHO 2016). The most recent WHO definition of infertility is the number of women of reproductive age (15–49 years) with a possibility of becoming pregnant (not pregnant, sexually active, not using contraception, and not lactating) who report trying for a pregnancy for two years or more. Other definitions refer to five years or more of desired parenthood and absence of pregnancy. Hence, infertility is usually recognized, "not by the presence of pathological symptoms, but by the absence of a desired state" (Greil et al. 2010: 141). Studies reveal that the overall pregnancy rate per cycle after IVF surrogacy was only 24%, with a clinical pregnancy rate of 19%, and a live birth rate of 15.8%, while the clinics usually publish exaggerated success rates on their websites (Goldfarb et al. 2000). Practitioners in Canada are saying that it is not only an increase in infertility but also the knowledge about IVF that is encouraging women to delay motherhood, in other words, technology is producing infertility (Kozicka 2016). Moreover IVF technologies and its possibilities are also increasing the social and subtle national pressure on women to use these methods for genetic lineage.

Postcolonial Surrogacy: Global North–South[1] Market Flow

The legalscape (drawn on Appadurai 1996 'global scapes') reveals a patchwork of countries in the global North that have well-drawn policies prohibiting or permitting surrogacy and effective implementation of these policies, while the countries of the global South struggle with the regulation of surrogacy biomarkets. However, the global North–South movement is inadequate in explaining the entire surrogacy ethnoscape. Similar patterns of people moving from affluent countries in Western Europe to not-so-affluent countries in Eastern Europe or Western Europe is also evident, forming complex constructions of global assemblages and bioavailability within the continent. The classism pattern becomes even clearer when affluent people from within countries use the services of poor people of that country regardless of whether it is Asia, Europe, or America. Hence, it is the one-third (haves) and two-third (have-nots) worlds that adequately describe the classism in the global surrogacy markets. Mohanty (2003) explains the advantage of one-third/two-thirds world over terms like western/third world and North/South as it moves away from misleading geographical and ideological binaries. However, this does not mean that the classic postcolonial pattern of flows from global North to global South is redundant or non-existent. The most lucrative returns still retain this classic flow of dollars/Euros into countries like India that specially attracts service and exploitation according to the demands of the intended parents. But even the one-third group within countries like India exploits people belonging to the two-thirds to satisfy their *Kinderwunsch*, desire for children. It is not just the intended parents who move to access surrogacy, surrogate mothers, gametes, and embryos. Even surrogate mothers were moved to different places in Asia following the prohibition of surrogacy in some countries. Frozen gametes and embryos also move from Europe to Asia and in the reverse direction to serve the needs of the intended parents.

I have focused on the more obvious, classic *globalscape* (as referred to by Appadurai 1996) of people moving from the global North (comparatively richer continents) to global South (poorer continents) for ARTs, especially for surrogacy and in the next section I have explored the globalscape within the global North, specifically Europe. Several writers emphasize that the global North–South boundaries are dissolving giving rise to a new pattern of South–South North–North movement for medical tourism. However, in the case of surrogacy the global legal patchwork, lower cost, and several other advantages in unregulated and ineffectively governed Asian countries made this continent an ideal biomaterial marketplace for people from countries prohibiting surrogacy or having well-defined rights for the surrogate mothers. Hence, the global North–South ethnoscape was more visible until several countries in Asia banned surrogacy for foreigners.

The movement from the global North to the global South is also evident in some African and South American countries. Mexico became a hub for people seeking

---

[1]I have referred to the market flow as Global North–South as it aptly describes the postcolonial pattern revealed in this section.

surrogacy from the USA. Similar stories of poor women desperate for money being involved in surrogacy began to emerge from Mexico following which the practice was banned there too. Instead Nigeria, Ghana, Argentina, Laos, Dubai, Iran, and Lebanon have developed into surrogacy hubs. Single mothers in Malaysia have been known to become surrogate mothers for childless couples from Singapore, although limited data exist on this. Thus, surrogacy has been described as the "hidden world" of assisted reproduction. This is so because the activity involves law evasion, stigma, secrecy, and anonymity. The legal and human rights conditions under which surrogacy occur are largely unknown in many of these countries.

The cross-border surrogacy movement clearly reveals a significant pattern and familiar spatial flow of people moving from the global North (developed countries) to the global South (developing countries) seeking surrogacy. To begin with, the comparatively more developed countries have a well-regulated and implemented legal system for surrogacy in place. Most countries in Europe including Austria, Denmark, Finland, France, Germany, Hungary, Iceland, Italy, Latvia, Lithuania, Norway, Poland, Slovenia, Switzerland, Spain, and Sweden prohibit all forms of surrogacy agreements. Yet, in other countries (Belgium, Denmark, Ireland, and the United Kingdom) only altruistic surrogacy is allowed. Although these countries allow surrogacy, there are direct or indirect restrictions on homosexual and singles aspiring to become parents using surrogacy. A well-drawn surrogacy policy and implementation in these countries also means more legal protection and rights for the surrogate mothers. While most countries of the global South are presently facing ineffective implementation of policies regardless of whether surrogacy is permitted or prohibited, intended parents from the global North prefer such unregulated markets where the surrogate mothers have no rights over the child or their own body, and owing to prevalence of unethical medical practice and ineffective governance they can undergo easy procedure and make a trouble-free exit from the destination country. Not all countries with unregulated surrogacy policies become a surrogacy hub; aspiring parents also seek high quality medical care, equipment, and facilities, fluent English-speaking medical practitioners and a comparatively lower cost (Crompton 2007).

India became one of the favourite global surrogacy destinations, providing standardized medical services and English-speaking doctors along with an unregulated market, ineffective governance, unethical medical practices, minimal rights of surrogate mothers, as well as a steady supply of socio-economically disadvantaged women willing to become surrogate mothers. Along with being legally deprived of any rights over their body or the baby, these women are also willing to abide by all the rules imposed by the clinic and the intended parents in their desperation to bring their families out of poverty (see Chap. 5). No upfront payment was required nor payment for miscarriage. Thus, this was a convenient option for many seeking surrogacy. This led to exploitation and several human rights violations which has become evident from the several studies emerging from India (SAMA 2012; Saravanan 2015, 2016). Given these conditions, many couples poured into India seeking surrogacy even though they belong to countries where the practice is permitted. It is estimated that 60–80% of all surrogacy pregnancies were commissioned by foreigners (Bhalla and Thapliyal 2013). According to Dr. Sudhir Ajja from a Mumbai-based fertility bank that has produced 295 surrogate babies, he catered to 90% overseas clients and 40% same-sex couples, since it opened in 2007 (Bhalla and Thapliyal 2013). Dr. Nayna Patel boasted of 500 babies in 2013, two-thirds of whom were for foreigners and people of Indian origin living from over 30 countries (Bhalla and Thapiyal 2013). According to Anupriya Patel, Minister of State, Health and Family Welfare in India, 80% of the surrogacy was catering to foreign couples (Ghosh 2016). Hence, surrogacy practice, just like the garment industry, is a specific form of classist and racist exploitation of reproductive labour directed at the vulnerabilities of women primarily from the Global south catering to demands in the Global North. Given this pattern, if this practice is legalized all over Europe, needless to say, it would be primarily the refugees and women of the lowest rung who will be exploited even in the Global north.

Previously, it seemed that India was the only prominent destination for global commercial surrogacy. When Nepal banned surrogacy for foreigners following the post-earthquake drama of evicting Israeli babies from the disaster zone, the inhuman approach in the surrogacy industry flourishing there became evident. A booming market in Thailand came to be known worldwide only after the media brought up a few controversial surrogacy cases. In 2014, Baby Gammy and Pipah were born to Pattaramon Chanbua through surrogacy commissioned by an Australian couple using the intended father's sperms and an egg donor. At a late-stage pregnancy, it was revealed through an ultrasound examination that one of the twins, Baby Gammy, had Down Syndrome and the intended parents requested Pattaramon to abort the baby. Pattaramon, on account of her Buddhist belief, chose to keep the pregnancy. On birth of the twins, Baby Gammy was born with a congenital heart condition along with Down Syndrome. The parents returned to Australia with Pipah while Baby Gammy remained with Pattaramon. She sought orders from Western Australia's Family Court to have Pipah returned to her which set off a legal struggle between the intended parents and the surrogate mother. According to the intended parents, on birth of the twins they wanted to take both the children back to Australia, but Pattaramon refused to give away Baby Gammy. She is said to have claimed the custody of both children based on her judgement that the intended parents were insensitive as evident from their decision of leaving behind a disabled child. It turned out later that the intended father had been a convicted sex offender on which the Australia's Department of Child Protection launched a separate investigation. The court decided that he had not re-offended any sex offence since his release from jail in 1999 and hence Pipah could continue to stay with the intended parents. The Australian court also stated that the couple had not abandoned Baby Gammy and it was Pattaramon who fell in love with the twins during the pregnancy (ABC 2016).

This case draws attention to certain ethical questions—rights of the surrogate mothers over her own body, particularly her opinion on aborting or keeping the child; child's rights to know about the surrogate mother; contradictions prevailing between notions of "*Wusnchkinder*" (wish child and the characteristics of such a child) and rights of a disabled child; and people with criminal background aspiring to be parents through transnational surrogacy. Some authors have justified that surrogacy is comparable to any other work or product in a labour market (Kirby 2014; Pande 2009). However, this case brings forth the complexity of surrogacy that makes it different from other forms of labour. Taking into consideration human sensitivities and the bonding that a surrogate mother can develop with the baby, surrogate mothers cannot be seen as baby-making machines nor can the babies be compared with machine-made products.

In 2014, another case involving a Japanese man who had fathered 16 babies through surrogacy in Thailand, came in the news as "the baby factory scandal". The Japanese man had given an online advertisement looking for surrogate mothers for a payment of $10,000. Women who were poor sought this as a solution to their poverty or to clear their family debts, and chose to become surrogate mothers without knowing that there was no couple but a single man. Eventually, the man went on to recruit 10 other surrogate mothers. On raiding a house in Bangkok, police found 16 babies (with nine nannies) fathered by this man. Twelve of the 16 children are being cared by social service organizations in Thailand. Some of the surrogate mothers were willing to raise the children if required. Although the man was charged with child trafficking motives, his objective seems to have been the desire for a big family. The market in Thailand had been thriving on poverty, secrecy, deception, and legal loopholes. As a result of these two cases, in February 2015, Thailand's parliament passed a law banning foreigners from seeking surrogacy in the country. The law bans non-Thai, homosexual, and single couples from accessing surrogacy in Thailand. Ongoing surrogacy contracts were allowed to return with their babies. However, during this period one surrogate mother refused to sign the adoption papers. After handing over the baby, she found out that the intended parents were a gay couple and she filed a legal notice asking for Baby Carmen to be returned to her (Sherlock 2016). Eventually, the intended parents won the custody case and Baby Carmen was handed over them. This couple already had a two-year-old son born in India through surrogacy. Surrogacy can also be motivated by frivolous activities and fears of pregnancy among some affluent women. One surrogate mother told me that her Indian intended mother was scared even of injections and she would not want to go through the process of pregnancy for fear, therefore commissioned a pregnancy with a surrogate mother in Ahmedabad. I am not trivializing the fear of pregnancy that some women may experience but this fear with a combination of geneticization, affluence to outsource pregnancy, and availability of markets that make such possibilities easily available brings the process dangerously close to being inhuman.

Meanwhile, in 2015, a massive earthquake hit the hill country of Nepal killing approximately 9,000 and injuring 22,000 people. Amidst this disaster, what also caught people's attention was the Israeli government's effort in rescuing 26 new-born babies born to Israeli parents through surrogacy along with Israeli trekkers and tourists stranded in Nepal. The government of Israel had sent rescue planes to airlift these babies leaving behind about 100, some heavily pregnant surrogate mothers in the disaster zone. Since Israel prohibits gay couples and single men from accessing surrogacy, India had been their favourite destination for several years, but due to a change in the law in 2013 prohibiting homosexual couples from accessing surrogacy in India, many contracts were caught midway through the surrogacy process. Many of these women from India were shifted to the neighbouring unregulated and unsuspecting Nepal to give birth to the children and some of the Israeli surrogacy agencies too shifted to Nepal. Local newspapers all over the world including in Israel critiqued the Israeli government for their selective empathy focusing only on the new-born infants and the plight of the Israeli citizens stuck in Nepal, ignoring the risk of women who had carried them (Green 2015; Kamin 2015).

At the time, the surrogacy law in Nepal was unclear regarding the rights and duties of the surrogate mothers; status and rights of the child(ren) delivered through surrogacy including baby's right to inherit parental property; the responsibility of fertility clinics; medical malpractices and violation of medical ethics. Further questions were also raised regarding statutory procedure to grant citizenship for the children and legal parenthood of the child(ren) born through surrogacy in Nepal. Although the earthquake accidentally exposed the surrogacy practice in Nepal, the industry had been mushrooming for several years with local reports of disputed inheritance rights of children born through surrogacy (ToI 2011). There were also reports that surrogate mothers were taken from Nepal to India to deliver out of fear of social stigma and critiques emerging from within Nepal (Subedi 2015). Women were exploited in Nepal by agents and family members to carry babies for foreigners in lieu of lucrative returns, and this was not openly reported or spoken about (Parajuli 2015). Soon after this incident, in August 2015, Nepal banned surrogacy for all foreigners. Following the ban, dozens of families spent weeks in limbo, unable to leave Nepal with their children. It took weeks of diplomatic negotiations before these children got their visas to exit the country. Many writers took up this case to advocate for equal reproductive rights for homosexuals while downplaying the double-standard attitude towards the surrogate mothers in India–Nepal (Maulem 2015; Conceive Abilities 2016). The case of Nepal raises ethical questions about how the movement of biomaterial caters to the needs of the global North as well as the affluent from within Nepal and the Asian subcontinent. Further there were other concerns with regard to medical malpractices, cheating (wives not informed about husband's involvement in contracting a surrogate child), child rights, surrogate mother's rights, inheritance rights, and health care issues of surrogate mothers and children, exploitation and other ethical questions raised in surrogacy practice in India, Thailand and similar other nations of the global South.

In October 2015, a documentary film group came to interview me in Goettingen University and the first question they asked me was; "what is your impression about the recent prohibition of surrogacy in India; wasn't the government profiting by this practice, is it anything to do with religion?" Those asking these questions were primarily searching for answers to complex questions in cultural relativism. I answered by drawing analogies between surrogacy and violence against women and replied with the question "do we ever doubt a law against any forms of violence against women?" This was immediately after the prohibition but it soon became clear that the prohibition of transnational surrogacy in India was not as sudden as it may have seemed to be but had come up after several considerations of court cases filed within India and consultations with medical practitioners, ministries, and stakeholders (Bhattacharyya 2016). The 228th report of The Law Commission of India published almost eight years ago has been a key driver in advocating for allocation of ethical altruistic surrogacy for only those couples of Indian citizenship who are in need of the service (GoI 2009; Bhattacharyya 2016). According to Mohan Rao, the recent regulation in India is also based on the HFEA (Human Fertilisation and Embryology Authority) guidelines of the UK (Kuchroo 2016). It was a chain of incidents and legal court cases not only in India but all over South Asia that led to the ban. India has witnessed some of the most difficult cross-border surrogacy situations and prolonged court cases. There have been several cases of death among surrogate mothers and egg donors, trafficking of teenage girls and women into surrogacy, single mothers illegally being exploited for surrogacy, abandonment of disabled children by intended parents that became evident through news media and research. My research conducted in just two clinics in Western India revealed several ongoing illegal surrogacy cases, near-death situations of surrogate mothers, neonatal and perinatal mortalities, unreported abandonment of disabled infants by intended parents and morbidities among surrogate mothers. Hence it is evident that the cases published in the media are just the tip of the iceberg, most of these cases go unreported. Rather, the Government of India has been somewhat slower to react in comparison to many other countries in South Asia that responded promptly to the emerging evidence of exploitation taking place in their respective countries. Before the ban, the criticism was that the regulation of transnational contracting of wombs in India is 'deliberately lax'. After the prohibition of surrogacy in India, the rhetoric that was glorifying the Indian women's wombs as gifts for global sisters turned into expressions of sympathy primarily for the western intended parents. A stream of media articles reflected both cultural imperialism and re-orientalism: 'India surrogacy ban dismays British couples', one article in *Telegraph*, 'Despair over ban in India's surrogacy hub' in *BBC*, 'India proposes commercial surrogacy; live-ins, homosexuals worst hit' in *Hindustan Times*, and 'Surrogacy Bill violative of privacy rights' (Sherwell 2015; BBC 2015; Gupta 2016). The sympathy towards western, homosexual and live-in couples and privacy rights seems to have taken over the concerns of human rights violation of the surrogate mothers.

Until the present Surrogacy Regulation Bill, 2016, was introduced by the Ministry of Health and Family Welfare, there was only a National Guidelines for Accreditation, Supervision and Regulation of ART clinics that was formulated in 2005 by the Indian Council of Medical Research (ICMR) and National Academy of Medical Sciences (NAMS). According to this guideline, the surrogate mother was not considered to be the legal mother of the child(ren) born though surrogacy and the birth certificate was issued in the name of the intended parents. Several other loopholes and lack of clarity in the guideline raised intense critique among non-governmental organizations (NGOs), academics, lawyers, medical practitioners, and media all over India (Qadeer 2010; Sarojini and Sharma 2009; Kusum 2015). Taking advantage of this ambiguity several ART clinics mushroomed all over

India, primarily around tourist areas (Deonandan et al. 2012). One clinic in Anand in Western India became very popular for its surrogate homes. The official registration of in vitro fertilization (IVF) clinics in India is inconsistent as records reveal that many clinics remain unregistered. According to a recent ICMR (Indian Council of Medical Research), 385 IVF registered clinics were operating in India (ICMR 2015). One of the most popular clinics in Western India claims the birth of more than 1,000 babies through surrogacy until 2016. According to a recent estimation, the surrogacy business in India accounts for a yearly return of 2.3 billion dollar (Perappadan 2014). Couples from abroad with infertility issues were specifically attracted to India for surrogacy due to the liberal laws that prevailed until 2016, ineffective law implementation that made it easier for people to find loopholes, low costs, easy availability of women willing to become surrogate mothers, fewer rights were granted to surrogate mothers in India compared to the USA or Canada, accessibility to English-speaking doctors, and standardised medical services. Surrogate mothers usually want to enhance their standard of living or savings to ensure that their children can receive higher education. Medical practitioners are in this business mainly for profit making and many other intermittent agents for earning a living. Surrogate mothers receiving a lower share of the surrogacy charge in India (15–25% of the surrogacy) as compared to the USA (35%) are able to buy a 10 m^2 house with the money they receive for the surrogacy (Saravanan 2015). However, the clinics have grown in size and capacity; this clinic in Anand after the success of delivering 1000 babies has expanded from a small clinic (Photo 2.1) into a multiplex sprawling approximately 22 square meters (Photo 2.2) with an all-inclusive facility of the IVF clinic, hotels for intended parents, surrogate homes, restaurants, theatre rooms, child hospitals, shopping areas within one complex costing millions of rupees.

In India, controversies began when Baby Manji born in July 2008 in Anand was stranded in India as the Japanese intended parents had divorced during the surrogacy (ToI 2008). According to Indian law, a birth certificate needs to have the father and mother's name. However, as the parents had divorced and Yamada's ex-wife was not the egg donor for the child, she had no responsibility towards the child, neither did the egg donor. The surrogate mother had signed off all the rights over the baby when she signed the surrogacy contract and she could not be handed over this responsibility. In this case, the father desperately wanted parental custody but according to Indian law, a single father cannot be handed over a baby even if he is genetically related to the child. Hence, Baby Manji's grandmother (father's mother) had to come to India and take custody of the child and return to Japan in November 2008 after the Japanese government issued a one-year visa to the baby on humanitarian grounds. As far as the clinic was concerned, their duty seemed to end by handing over the baby. This case received a wide media coverage and initiated an intense public debate on surrogacy in India.

Nikolas and Leonard Balaz, also born in Anand, in 2008, were another significant case as the twins were stranded in India, stateless for several years. The father stayed in India to fight for the custody of the children, arguing for an Indian

Photo 2.1 A surrogate home in Gujarat. *Source* Author, 12 September 2009

Photo 2.2 The new all-inclusive multiplex clinic in Gujarat

citizenship to be given to them. Initially, Indian passports were issued to the children and it was only due to an immigration enquiry from another country in Eastern Europe that the German and Indian consulate realized that this was a case of surrogacy and their passports were recalled. The clinic played a major role in persuading the intended parents into believing that they could evade the law in India by convincing them that several Germans have thus returned with their babies. I met one intended mother from Germany living underground with her children in India and heard of another German man who had signed a surrogacy contract with a single mother in Mumbai. I also came across several other couples who had come from countries that prohibited surrogacy (Netherlands, Denmark, Australia). Another case that was exposed was of the twins born through surrogacy in 2009 in India to a single mother, Andras Bell from Norway and remained stateless because she was genetically unrelated to the children as she had chosen a sperm donor of Scandinavian origin and an Indian egg donor (Roy 2010). In 2010, Liron and Itai were born through surrogacy also in Mumbai, and the intended father Dan Goldberg had to undergo a paternity test for them to get Israeli citizenship (ToI 2010). However, the Jerusalem family court refused to allow a paternity test to initiate the process for Israeli citizenship as it was not under Israeli jurisdiction. It was only when the prime minister intervened following protests by the Left wing and the gay community in Israel that a paternity test was initiated and the twins were given Israeli citizenship. Due to such unresolved cases, in July 2010, eight countries who had banned surrogacy, took an initiative to issue letters to all the registered clinics in India through their consulates to refrain from providing surrogacy services to their citizens.

In these cases, conditional nationality registration would infringe upon the child rights. According to the Convention on the Rights of the Child, any child has the right to be registered soon after birth with a nationality, name and family relations, to freely enter one's country with parents and not to be subjected to arbitrary or unlawful interference with his or her privacy and family. Anger in such cases is thus directed towards India as a nation holding the primary responsibility to provide citizenship to the children rejected by the source countries because these children were indeed born in India. Germany is considered having lesser responsibility towards the children as the citizens of their country had been involved in a practice considered illegal within their borders. In all cases, eventually the source country along with the destination country have mediated the process by providing adoption-parental rights to the intended parents.

One of the hushed-up cases was of the twins, a girl and a boy, born through surrogacy in India to an Australian intended couple in November 2012. The intended couple was adamant to abandon the boy child and return only with the girl because they could not afford him. They already had a son at home and wanted to "complete their family" with a girl, according to the communication between the Australia consulate and the couple (Hawley and Smith 2015). They were repeatedly told that abandoning the boy could leave the child stateless because India did not recognize surrogate children as citizens. They insisted of having followed the Indian law in handing over the child for adoption in India. This misled the Australian consulate who understood that the couple had made arrangements for handing over the baby boy to a friend of theirs. There are reports that money had been exchanged in the handing over of the baby to another couple in India. The Australian couple had sold the baby boy in India which is analogous to child trafficking.

Other cases which are kept suppressed are those of the surrogate mothers and egg donors who have died in India. In May 2009, a 23-year-old girl named Easwari died of severe post-partum hemorrhage after giving birth (Global Bioethics Blog 2012).

She was referred to another hospital for treatment as the clinic was ill-equipped and was also asked to pay for her own transport expenses; she died en-route. Easwari was the second wife in a polygamous marriage and the husband had seen an advertisement in a local newspaper and coerced her into this process. The baby was handed over to the intended couple and the clinic denied any responsibility in the event. In 2010, 17-year-old Sushma Pandey (not an adult according to Indian Law) died in Mumbai after the egg extraction procedure. She would earn $448 for egg donation, more than what she earned working in a garment shop, hence, she opted to become an egg donor to help the family financially (Chatterjee and Janwalkar 2014). The clinic where she died is the workplace of one of the most prominent doctors providing surrogacy services in India. The case is ongoing in the court and the doctors claim that Ms. Pandey had been to the clinic several times for egg donation and they claim that even if there had been a complication, it must have not been due to some "extraneous reason" which occurred outside the hospital and it was not the responsibility of the doctors. Premila Vaghela a 30-year-old surrogate mother in India, died in a clinic in Gujarat, after suffering a convulsion (ToI 2012). The baby was meanwhile removed through caesarean section as she showed signs of life-threatening health distress. The hospital where she was transferred to reported that she died of a severe cardiac arrest and had been brought to them in a critical condition. Surrogate mothers must sign in the contract that they take full self-responsibility in case of any adversity or health disaster. Hence, the extra compensation of Rs. 10 lakh that was paid to her family was considered adequate. She was just like most surrogate mothers who are attracted to surrogacy due to their poor socio-economic background. Twenty-three-year-old Yuma Sherpa, died of ovarian hyper-stimulation syndrome in a south Delhi fertility clinic on 29 January 2014 while undergoing a procedure for retrieval of eggs. Her husband alleges that Sherpa was persuaded by agents of the clinic to become a donor. According to him, although she complained of discomfort during the egg retrieval procedure, they compelled her to continue and she died. Her lawyer said, clinics were harvesting many more eggs than was recommended, sometimes up to 50 (Indian Express 2016). What goes unreported though is the near-death situations and those causing life-long health or psychological problems. The other unknown health impacts that surrogate mothers face is from the excessive and repeated doses of hormones as many have to go through several trials before a successful conception.

These reported deaths of surrogate mothers and egg donors make up the tip of the iceberg. Drawing analogies with reported cases of violence against women in India, mortality implies several cases of morbidities that go unreported. Moreover, several cases of missing girls and trafficking has been accounted to commercial surrogacy in India. Young girls were reported to have been kidnapped from remote areas of India (North–East, Jharkhand, and Chhattisgarh) and transported to metropolitan areas and coerced into surrogacy practices (Roy 2015; Bhattacharyya 2016; Nair and Sen 2004). In February 2015, there was news of a 13-year-old girl who was trafficked from Jharkhand into Delhi, forced into surrogacy and made to deliver six children. She was made to breastfeed the babies for six months before they were sold. "They treated me like a money minting machine. My will never mattered to them, all they wanted was me to deliver babies for them" (Roy 2015). She had no idea where her children were sold and to whom and is now seeking justice from Child Welfare Committee in Gumla. In the same district in East India, another girl was trafficked into surrogacy when she was just eight years old and forced to deliver at least 10 babies (Roy 2015). There are other cases of trafficking too from the same district. Some of the girls gave birth to children while staying in their houses and the agents came later to collect the babies. According to this report, some were paid a pittance and others were cheated after handing over the baby.

In India, surrogacy was allowed since 2005 and yet exploitation and commodification of women was showing an increasing trend. Most scholars who advocate for more regulation than a prohibition accept that surrogacy is exploitative. One point of contention is that surrogacy is difficult to regulate in a country as large as India, hence, a prohibition is likely to drive the practice underground. This is similar to saying that we cannot stop violence against women in India, hence it should not be criminalized. Prostitution is permitted and supposedly regulated in India, but exploitation and trafficking is only increasing. According to the National Crime Record Bureau, India, almost 20,000 women and children were victims of human trafficking in 2016, a rise of nearly 25% from the previous year (ToI 2017). In recent times, there are suggestions that low cost infertility services need to be provided to the socio-economically disadvantaged people in countries like India (Pennings 2008; Sallam 2008). The question that needs to be posed though is whether provision of affordable infertility choices to the poor can eradicate exploitation and alienation underlying reproductive stratification in practices like surrogacy (Roberts 2009). This liberal solution of providing free and easy individualized access to IVF facilities to all in the same tone of 'health for all' as a constitutional right reinforces several forms of social prejudices.

Considering all these cases, in September 2016, the Union Cabinet approved the Surrogacy (Regulation) Bill, 2016, according to which, commercial surrogacy is completely prohibited and foreigners cannot access surrogacy in India, but altruistic surrogacy is permitted only for married couples in need with the help of close relatives as surrogate mothers. Violators would face at least five years in jail and a fine of up to INR1 million (nearly $15,000) (Zargar 2016). Surrogate mothers will have more rights over the child and will be offered legal support. In support of the bill, external affairs minister Sushma Swaraj said it will protect women from exploitation, especially by the rampant medical tourism industry. Anupriya Patel, Minister of State, Health and Family Welfare questioned, "Is she a child-producing factory? Families often coerce women into surrogacy due to poverty. Should it not be stopped?" and rightly so (Ghosh 2016). From the neoliberal perspective, the use of technology to realize the intended parent's "reproductive goals" is considered a constitutional right, hence, the state's intervention is seen as interference, some have even gone to the extent of calling it state policing on people's private life. This individualistic, outcome-based approach overlooks the social impact of surrogacy and the structural injustice, racial and colonial elements of this industry.

The government took years to respond effectively but has finally taken a commendable stand with an understanding that this entire industry is based on structural inequalities, exploitation and commodification of women and children. India prohibits surrogacy to certain groups of people (homosexuals, single parents, live-in couples). This is a partial solution to the problem, as it was only the affluent people from various sections of the society who could afford surrogacy in any case. This policy is indeed patriarchal and homophobic and fundamentally discriminates against some sections of the society. For example, Germany restricts gamete donation for unmarried and homosexual couples and for people who are carriers of genetically inheriting conditions. Such selective restrictions promote conservative notions of family formation and contradict the constitutional rights to privacy, health, and freedom from discrimination (Krause and Marchesi 2007).

After the prohibition of surrogacy in India, the police raided an illegal fertility clinic in Telangana in June 2017 to find 47 surrogate mothers inside a house, mostly brought in from the north-eastern region of India to rent their wombs in return of money. The official from the clinic who was interrogated said, they were operating within the law as the women were not confined against their wishes (Nagaraj 2017). The northeastern region in India has been targeted by traffickers for prostitution, marriage, and bonded labour trade. Kailash Satyarthi, an anti-human trafficking and child labour activist said, "Children from the North–East, mostly minor girls, are trafficked for being used as domestic helps in metros and are physically abused and sexually exploited as well. Every year, 4,000 children go missing from the state." Vulnerable girls were trafficked by luring them for enhanced income source into prostitution and bonded labour and now surrogacy has added to this income source list (Baruah 2012).

The surrogacy industry, which catered to foreigners, abruptly halted its operations in India, Nepal, and Thailand, and has shifted to Cambodia (Bhowmick 2016). According to estimates 50 infertility clinics moved to Cambodia after it was banned in South Asian countries and this was the primary reason for the ban on surrogacy in Cambodia in November 2016. There were reports that many women recruited as surrogate mothers in Cambodia were Thai women. As there was no clear law on surrogacy in Cambodia, intended parents have been warned by international agents against engaging in surrogacy contracts in this country (Murdoch 2015). Australians were known to be the largest buyers of surrogacy services in Cambodia. Surrogate mothers have been physically transported from India to Nepal, as well as from Thailand to Cambodia, following the movement of the market. With the prohibition of surrogacy to foreigners in all these countries, Nigeria, Argentina, Brazil, Laos, Dubai, Iran, and Lebanon have developed into surrogacy hubs. In Nigeria and Brazil, there are no laws prescribing or proscribing surrogacy, creating a void, which is being and can be exploited further by illegal commercial operators. In 2013, 14 pregnant teenage girls were rescued from a baby-making factory in Umuguma, Nigeria. A pastor recently rented an uncompleted building in the community and lodged the pregnant girls in an unhealthy environment. In another incident at Umuka, the police raided a baby factory to rescue 26 pregnant teenage girls along with the arrest of a 23-year-old boy who was hired to impregnate the girls in a child trafficking case (Nkwopara 2013). In Ghana, surrogacy practice is gaining grounds in some parts of the country with Accra being the main hotspot. It is largely the commercial sex workers who are willingly participating in the industry. The number of babies born is estimated to be in hundreds. There is not much information on surrogacy in Ghana except for a few cases of cheating. A surrogate mother carrying quadruplets was cheated with incentives, no contract was signed between her and the IVF clinic (Daisie 2015). There have been few studies on the health or psychological impacts of surrogacy on the surrogate mothers, the children or on the intended parents in UK. There is one study conducted in UK in which children born through surrogacy responded to online questionnaire of not being unhappy with the surrogacy regardless of whether the surrogate mother had used her egg or not and almost half of the children were in contact with the surrogate mothers (Jadva and Imrie 2013). This cannot be compared to the situation in India because surrogate mothers in India are mostly not asked about their preferred relationship with the child. Another study found that the relation between the intended mother and the child born through surrogacy showed less positive interaction than the natural conception families (Golombok et al. 2011). This was the reproscape in the global South, but a closer look at the continents reveal a similar transnational reproscape of movement pattern within Europe, from more developed and wealthier countries in the West to those in the East and South.

Europeanization of Surrogacy Markets

Europe has experienced new modes of spatialization as a reaction to challenges posed by recent socio-political transformations also termed as 'Europeanization' (Fergusson and Gupta 2002). Eastern Europe is known for offering lower prices while providing Western standards of IVF services (Knoll 2012). Patterns of these new regional nodes of assisted reproductive technologies is evident in Europe; the Germans move to the Czech Republic for egg donation, the Dutch and French move to Belgium for sperm donation, the Italians move to Spain, and the Swedish and Norwegians move to Denmark. Spain, Belgium, and the Czech Republic are known to be the main destination countries for pre-implantation genetic diagnosis (PGD) (Corveleyn et al. 2007). Aspiring parents from Germany have largely sought PGD in the Czech Republic, Benelux countries, Western and Southern Europe. Denmark, Belgium, and Spain are the main destinations for sperm donation and the Czech Republic and Spain for egg donation. People from Germany are known to prefer gamete donors from Eastern Europe (Poland, Czech Republic) rather than Spain for similarity in phenotype and although they may visit Spain for the medical process, they prefer gametes to be imported from Eastern European countries. Germans seek egg donation in the Czech Republic and Spain, and sperm donation in Denmark and Belgium (Präg and Mills 2015; ESHRE 2008; Pennings et al. 2009; Corveleyn et al. 2007). Popular locations for commercial surrogacy in Europe are Ukraine and Russia. Commissioning parents from countries in Western Europe move to Eastern European countries such as Ukraine, Poland, Romania, and Republic of Georgia for surrogacy.

As Spain has become a hotspot for trans-European oocyte donation, it is interesting to review how this phenomenon is naturalized and constructed by gamete seekers. A study on Danish women and couples moving to Spain for egg donation perceived oocytes as spare parts and exotic substances as well as forms of new kinships through shared blood, space and desire of shared femininity (Kroløkke 2014). In the case study, from the perspective of the recipients of gametes, the Danish women travelling to Spain seeking egg donation, "Western-ness" becomes synonymous with quality and whiteness compared to Eastern-ness (Kroløkke 2014). Another study revealed that German's travelling to Spain for oocyte donation seek the donors from Eastern Europe as per desired phenotypes (Bergmann 2011). Some Germans seek egg donation directly in the Czech Republic to be able to find oocyte donors with blue eyes. These desires call for more attention to racism, sexism, gendered, nationalist biases, not to mention ableist prejudices in this industry (Nahman 2008). Such particular demands by gamete seekers has generated a possible gendered part-time job for migrants in Spain while in the Czech Republic donors are recruited from poorer rural regions (Bergmann 2011). According to demands, clinics in Spain recruited East European migrants or Erasmus students as donors to provide the northern European oocyte seekers with their desired phenotypes (Bergmann 2011). A clinic in Spain asks migrant white Brazilian egg donors to outline three generations of 'whiteness' to reduce chances of racial mixing (Kroløkke 2014). Clinics/agents have detailed information regarding egg donors from which the gamete seekers may choose. The most worrying concern is how precisely these demands are normalized, naturalized, and necessitated. It is slowly turning into a bazaar in which compassion, authenticity, empathy, and justice is turning into cruelty, dishonesty, apathy, and oppression. This study lacks the perspective of the donors, but the gamete seekers understand oocyte donation as a form of giving, the donation as an act of gifting.

In January 2016, the European Parliament (EP) adopted a very important resolution on human rights and democracy in the world, where it calls for EU action on a series of issues including surrogacy. In this resolution, the EP "condemns the practice of surrogacy, which undermines the human dignity of the woman since her body and its reproductive functions are used as a commodity; considers that the practice of gestational surrogacy which involves reproductive exploitation and use of the human body for financial or other gain, in particular in the case of vulnerable women in developing countries, shall be prohibited and treated as a matter of urgency in human rights instruments".

This resolution came up after several incidents and legal controversies in Europe. Following the ban of surrogacy in Sweden, a journalist wrote, "India and Thailand do not want their female citizens to become the baby factories of the world. Now it is time for Europe to take responsibility. We need to show solidarity and stop this industry while we can" (Ekman 2016).

As I mentioned earlier most countries in Europe prohibit all forms of surrogacy agreements (Austria, Denmark, Finland, France, Germany, Hungary, Iceland, Italy, Latvia, Lithuania, Norway, Poland, Slovenia, Switzerland, Spain, and Sweden) and others allow only altruistic surrogacy (Belgium, Denmark, Ireland, and the United

Kingdom). The countries that permit altruistic frameworks have a well-drawn policy with legal protection and rights for the surrogate mothers. In Europe, popular locations for transnational commercial surrogacy are Georgia, Ukraine, and Russia. Commissioning parents from countries in Western Europe move to Eastern European countries, such as Ukraine, for surrogacy. Since many Asian countries banned surrogacy for foreigners, surrogacy agents have redirected their focus to countries such as Poland, Romania, and the Republic of Georgia. Although it is known that surrogacy is growing in many countries, there is very little information of the extent of this activity in Malaysia, Ukraine, Poland, Georgia, and Russia.

In the Czech Republic, there is no surrogacy law, hence it is neither permitted nor prohibited. According to a news report in August 2015, a baby born to a surrogate mother in the Czech Republic was abandoned due to its disability (congenital defects, including epilepsy and paralysis) and has been placed in institutional care (Prague Daily Monitor 2015). The baby was wanted neither by the biological parents nor the surrogate mother. Usually, the biological father is recognized as the parent, while the biological mother has to apply for adoption. The surrogate mother in this case was a single person suffering from epilepsy and had to give the child for adoption. Several factors led to the surrogate mother giving birth to the child and both the biological father and the intended mother consented to placing the child in a children's home after he underwent a series of surgeries soon after birth. This was dismissed as an unfortunate event by the Czech Gynaecological and Obstetrical Society Deputy Head, Vladimir Dvorak.

A lady from Bern, Switzerland, went to Georgia for surrogacy and the baby she returned with had to be taken into protective custody because her adult son, also one of the caretakers of this baby, was a convicted sex offender (Douez 2011). People can choose a surrogate mother and gamete donor by browsing through a list with photographs and details regarding age and race. Similar to concerns regarding surrogate mothers in the global South, women here too often have few employment opportunities and surrogacy provides them with a lump sum of money. In return, the surrogate mothers have to give up all rights over the child, even if she is the egg donor. Local authorities give a birth certificate in the name of the intended parents. In all terms, the woman is considered a "surrogate", in actual terms a proxy in the process. Women in less-fortunate situations are exploited for someone else's benefit. It's an alienated labour with predetermined detachment between the mother and the baby she carries. Here again, the context of consent is questioned when their financial situation is precarious. The intended parents are given a choice to keep their identity hidden from the surrogate mother, therefore, she cannot know the whereabouts of the child or what happens thereafter in the child's life. Given this context, Cora Graf-Gaiser Deputy Director of the Federal Office for Civil Status told swissinfo.ch that it should not be assumed that all who travel into Switzerland along with babies whose birth certificates contain their names will be accepted. The Federal Court emphasized that a child must be protected from being a commodity and observes that a pregnant woman cannot actively abandon her rights over the child before it is born.

In 2011, one commissioning couple from France was caught in their frantic attempt to illegally take their twin daughters by road out of Ukraine. These children were born in Ukraine to a surrogate mother under a surrogacy contract with a French couple. The homosexual Belgian couple struggled for two years before they could bring home their baby, born to a Ukrainian surrogate mother. Although rarer, a case of reverse exploitation has also emerged from France, of the surrogate mother cheating the intended parents. In 2010 after agreeing to a surrogacy contract for money, she informed the couple that the child was born dead while actually selling the child to another couple. In 2012, she repeated this crime with another couple and on interrogation testified that she had been sexually abused by her father as a teenager and suffered for a major emotional neglect and low self-esteem (The Guardian 2016).

Cross-border assisted reproduction is on the rise due to differential regulations, cost factor, differing accessibility and personal choice. In Europe, legal reasons were predominant for people crossing borders from Italy (70.6%), Germany (80.2%), France (64.5%), and Norway (71.6%). Aspiring parents moved from Italy to neighbouring Spain and Switzerland for sperm and egg donation, Germans move to the Czech Republic for egg donation, the British visit the Czech Republic and Spain also for egg donation, the Dutch and French go to Belgium mainly for sperm donation, the Swedish and Norwegians go to neighbouring Denmark for sperm donation. USA is a destination for people from Latin America, Europe, and Canada seeking anonymous egg donation. People seeking medically assisted fertility options move from restrictive countries to comparatively permissive countries. For these individuals, one of the most important reasons for moving between countries is restrictive legislation. A study in Belgium revealed that 56% of those who traveled for medically assisted reproduction did so for legal reasons. The other reasons include resource considerations (cost, personnel and equipment, and waiting lists), quality and safety concerns (quality of care and success rates), and personal preferences (cultural factors, support networks, and privacy concerns) (Inhorn and Patrizio 2012). Americans are known to seek reproductive services in South Africa, Israel, Italy, Germany, and Canada, where the costs are lower.

One of the legal challenges of global surrogacy is its prohibition in some countries and to certain groups of people, about which questions of political pluralism has been raised. Extra-territoriality brings forth the question of rights of citizens and children. Some of the suggestions put forward to reduce global and national inequalities are harm reduction and regulatory harmonization. These solutions continue to be debated both between and within disciplines, medical practitioners and policymakers. Extra-territoriality laws put restrictions on citizens who travel for fertility treatments and are not permitted to enter with their children born through surrogacy back into the source countries. Regulations are in place in many countries but the implementation of these laws is weak. In Germany, legal parental rights can be withheld for 2–3 years, while some other countries have imposed fines. However, most court cases grant parenthood and citizenship taking into consideration the best interests of the child. The law in the destination country decides about the citizenship to children born through surrogacy outside their jurisdictions and the parentage of the individuals who have travelled abroad to have these children. The countries which do not have clear extra-territoriality laws have been criticised for protecting their own citizens while allowing vulnerable citizens from other countries to be exploited.

The more affluent source countries have been criticized for allowing their citizens to avail surrogacy services by using the vulnerability of impoverished citizens in 'transitional economies' like India, while protecting their own citizens. The destination countries have also been blamed for viewing surrogacy merely as an opportunity for economic gains and thereby permitting objectification and exploitation of their own citizens, especially when the citizens of their own country do not have adequate access to basic health care services. While high quality reproductive health care is provided to surrogate mothers during their contract pregnancies, they have had almost no access to any quality health care for their own pregnancies. This disparity brings to light reproductive injustice that accords a higher value to certain pregnancies and babies (Bailey 2011). It is important to overview the kind of existing inequalities in transitional economies like India to understand the socio-economic circumstances in which the contract mothers make choices and decisions.

Conclusion

Globalization and developments in information and medical technologies has contributed to the easy cross-border access and transnational movement for ARTs. Exploitation of structural inequalities between the rich and poor and the global North and South is becoming a largely normalized phenomenon in the postcolonial era. A similar pattern of normalization is observed also within Europe, typically between unequal countries and between the affluent intended parents and poorer surrogate mothers. The surrogacy biomarkets reveal a clear postcolonial pattern where imperialistic intended parents have their wishes and desires fulfilled in India; parallel to this are movements from Western Europe to the lesser developed Eastern European nations. Some scholars would say that the global pattern observed in this chapter is neocolonialism alongside postcolonialism. Chapter 6 describes the postcolonial dilemma that combines neoliberalism and postcolonialism. In plain language, Nkrumah (1966), Young (1995), and Bhabha (1994) describe neocolonialism broadly as a continued master–slave relationship between the colonizers and colonized. Young (2003) observes that colonialism continues in a different form, power has transferred from the colonizers to the government controlled by a local elite class, the bourgeoisie, who are foreign qualified and less inclined to the promotion of local culture and political independence, giving a sense of false freedom. Hence, colonizers were replaced by corrupt leaders and feudal landowners (Sartre 2001). This pattern will be revisited in Chap. 5 in the description of the surrogacy bazaar in India.

These global reproscapes reveal the downside of the naturalization of geneticization that prioritize genetic linkage and devalues the surrogate mother's gestational role. While the medical evidence reveals that there is a two-way genetic transfer between the surrogate mothers and the baby, DNA traces of the first child is also found in the second child transferred through gestation which puts a question mark surrounding the surrogate mother's role. The consequence of this devaluation is also observed in the glorifying rhetoric of the altruistic motive of surrogate mothers and the normalization of alienating the woman's body from her womb which is discussed in Chaps. 3, 4, and 5 in the feminist discourse and the ethnography of the surrogate mothers and intended parents and role of the medical practitioners. The normalization of alienated reproductive labour also makes it easier for medical practitioners to rationalize surrogacy as any other manual work. Genetic reductionism has led to ableism, leading to abandonment of disabled children. It has also given rise to donor pools in various places of the world deeming some genetics superior and desired and others as undesired; this leads to price tags on highly desired human characteristics which are then normalized as being superior. Simultaneously, a necessitation of surrogacy as a solution to infertility, bases itself on the pronatalist definitions of parenthood. An advocacy for geneticization that reinforces the concepts of race and ableism encourages social differences and domination. This book examines surrogacy biomarkets in India by applying reproductive justice as a methodological framework from a feminist perspective as a pathway leading to transnational feminism and humanitarian assisted conception which transcends such social prejudices.

References

ABC. 2016. Baby Gammy: Surrogacy row family cleared of abandoning child with Down syndrome in Thailand. *ABC News*. http://www.abc.net.au/news/2016-04-14/baby-gammy-twin-must-remain-with-family-wa-court-rules/7326196. Accessed 14 Apr 2017.

Agamben, Giorgio. 1998. *Homo sacer: Sovereign power and bare life*. Translated by D. Heller-Roazen. Stanford: Stanford University Press.

Alpers, J.S., and J. Beckwith. 1993. Genetic fatalism and social policy: The implications of behavior genetics research. *The Yale Journal of Biology and Medicine* 66: 511–524.

Anderson, Elizabeth S. 1990. Is women's labor a commodity? *Philosophy & Public Affairs* 19 (1): 71–92.

Andrews, Lori B. 1988. Surrogate motherhood: The challenge for feminists. *Journal of Law, Medicine & Ethics* 16 (1–2): 72–80.

Appadurai, Arjun. 1996. *Modernity at large: Cultural dimensions of globalization*. Minneapolis: University of Minnesota Press.

Bailey, Alison. 2011. Reconceiving surrogacy: Toward a reproductive justice account of Indian surrogacy. *Hypatia* 26: 715–741.

Baklinski, Pete. 2014. Shock: UK mom abandons disabled daughters, keeps healthy son after twin surrogacy. *Life Site News*. https://www.lifesitenews.com/news/mother-rejects-disabled-daughter-keeps-healthy-son-after-twin-surrogacy. Accessed 14 Aug 2017.

Baruahl, Pranjal. 2012. Traffickers target northeast Indian women, sell them like cattle. *Times of India*. http://timesofindia.indiatimes.com/india/Traffickers-target-northeast-Indian-women-sell-them-like-cattle/articleshow/17725299.cms. Accessed 14 Aug 2017.

BBC. 2015. Despair over ban in India's surrogacy hub. *British Broad Casting*. http://www.bbc.com/news/world-asia-india-34876458. Accessed 14 Aug 2017.

Benninghaus, Christina. 2012. Beyond constructivism? Gender medicine and the early history of sperm analysis, Germany 1870–1900. *Gender & History* 24 (3): 647–676.

Bergmann, Sven. 2011. Reproductive agency and projects: Germans searching for egg donation in Spain and the Czech Republic. *Reproductive Biomedicine Online* 23 (5): 600–608.

Bhabha, Homi K. 1994. *The Location of Culture*. New York: Routledge.

Bhalla, Nita, and Mansi Thapiyal. 2013. India seeks to regulate its booming 'rent-a-womb' industry. *Reuters*. http://www.reuters.com/article/us-india-surrogates-idUSBRE98T07F20130930. Accessed 14 Aug 2017.

Bhattacharyya, Ritupurna. 2016. Draft surrogacy (regulation) bill 2016: Rhetoric of surrogate-centric. *Space and Culture, India* 4 (2): 9–21.

Bhowmick, Niranjana. 2016. After Nepal, Indian surrogacy clinics move to Cambodia. Aljazeera. http://www.aljazeera.com/indepth/features/2016/06/nepal-indian-surrogacy-clinics-move-cambodia-160614112517994.html. Accessed 14 Aug 2017.

Campbell, Mary. 2007. Thinking outside the (black) box: Measuring black and multiracial identification on surveys. *Social Science Research* 36 (3): 921–944.

Chatterjee, Pritha, and Mayura Janwalkar. 2014. The great Indian egg bazaar. *Indian Express*. http://indianexpress.com/article/india/india-others/the-great-indian-egg-bazaar/. Accessed 15 Aug 2017.

Corveleyn, A., E. Zika, M. Morris, E. Dequeker, J.L. Davies, K. Sermon, G. Antiñolo, A. Schmutzler, J. Vanecek, F. Palau, and D. Ibarreta. 2007. Pre-implantation genetic diagnosis in Europe. European Commission Joint Research Centre, Institute for Prospective Technological Studies. EUR 22764 EN–Seville: European Communities.

Conceive Abilities. 2016. Nepal joins India and Thailand in commercial surrogacy ban. *Conceive Abilities*. https://www.conceiveabilities.com/about/blog/nepal-joins-india-and-thailand-in-commercial-surrogacy-ban. Accessed 14 Aug 2017.

Corea, Gena. 1985. *The mother machine: Reproductive technologies from artificial insemination to artificial wombs*. New York: Harper and Row.

Cornell, Drucilla. 1998. *At the heart of freedom: Feminism, sex, and equality*. Princeton: Princeton University Press.

Crompton, S. 2007. *India's rise in medical tourism*. London, Britain: The Times of India.

Daisie, Maame Aba. 2015. Surrogate mother carrying quadruplets cheated of incentives. *My Joy Online*. http://www.myjoyonline.com/news/2015/January-12th/surrogate-mother-carrying-quadruplets-cheated-of-incentives.php. Accessed 14 Aug 2017.

Dawe, Gavin, Xiao Wei Tan, and Zhi-Cheng Xiao. 2007. Cell migration from baby to mother. *Cell Adhesion And Migration* 1 (1): 19–27.

Deonandan, Raywat, Mirhad Loncar, Prinon Rahman, and Sabrina Omar. 2012. Measuring reproductive tourism through an analysis of Indian ART clinic Websites. *International Journal of General Medicine*: 763

Dhar, RajibLochan. 2013. Lived experiences of childless couples: A phenomenological study from the Indian rural context. *Marriage & Family Review* 49 (4): 265–283.

Donchin, Anne. 1996. Feminist critiques of new fertility technologies: implications for social policy. *The Journal of Medicine and Philosophy* 21 (5): 475–498.

Douez, Sophie. 2011. Flourishing surrogacy business raises fears. *Swissinfo*. https://www.swissinfo.ch/eng/flourishing-surrogacy-business-raises-fears/29791340. Accessed 14 Aug 2017.

Dreyfuss and Nelkin. 1992. The Jurisprudence of genetics. *Vanderbilt Law Review* 45: 313–348.

Duster, Troy. 1990. *Backdoor to eugenics*. New York: Routledge.

Ekman, Kajsa Ekis. 2016. *The Guardian*. https://www.theguardian.com/commentisfree/2016/feb/25/surrogacy-sweden-ban. Accessed 15 Aug 2017.

European Society of Human Reproduction and Embryology. 2008. Comparative analysis of medically assisted reproduction in the EU: Regulation and technologies. (SANCO/2008/C6/051). *European Society of Human Reproduction and Embryology (ESHRE)*. https://ec.europa.eu/health//sites/health/files/blood_tissues_organs/docs/study_eshre_en.pdf. Accessed 14 Aug 2017.

Ferguson, James, and Akhil Gupta. 2002. Toward an ethnography of neoliberal governmentality. *American Ethnologist* 29 (4): 981–1002.

Field, Martha A. 1988. *Contract motherhood: The legal and human issues*. Cambridge: Harvard University Press.

Ghosh, Abantika. 2016. Surrogacy legislation: Is woman a child-producing factory asks Anupriya Patel. *Indian Express*. http://indianexpress.com/article/india/india-news-india/surrogacy-ban-commercial-foreign-clients-woman-child-producing-machine-3008933. Accessed 14 Aug 2014.

Gilroy, Paul. 2000. *Against race: Imagining political culture beyond the color line*. Cambridge: Harvard University Press.

Gingrich, Andre, and Gerd Baumann. 2004. *Grammars of identity: A structural approach*. New York: Berghahn.

Global Bioethics Blog. 2012. Contract motherhood in developing countries: Fine in theory, nasty in practice. *Global Bioethics Blog*. globalbioethics.blogspot.com/2012_08_01_archive.html. Accessed 14 Aug 2017.

Goldfarb, James M., Cynthia Austin, Barry Peskin, Hannah Lisbona, Nina Desai, and J.Ricardo de Mola. 2000. Fifteen years' experience with an in-vitro fertilization surrogate gestational pregnancy programme. *Human Reproduction* 15: 1075–1078.

Golombok, Susan, Jennifer Readings, and Vasanti Jadva. 2011. Families created through surrogacy: Mother-child relationships and children's psychological adjustment at age. Development Psychology.

Government of India. 2009. Need for legislation to regulate assisted reproductive technology clinics as well as rights and obligations of parties to a surrogacy, Report Number 228. Law Commission of India: New Delhi.

Greil, A.L. 1991. *Not yet pregnant: Infertile couples in contemporary America*. New Brunswick: Rutgers University Press.

Green, Alon-Lee. 2015. Where is the concern for the surrogate mothers in Nepal? *Haaretz*. http://www.haaretz.com/opinion/.premium-1.653660. Accessed 14 Aug 2017.

Greil, A.L., K. Slauson-Blevins, and J. McQuillan. 2010. The experience of infertility: A review of recent literature. *Sociology of Health & Illness* 32: 140–162.

Gupta, Moushumi Das. 2016. India proposes commercial surrogacy ban; live-ins, homosexuals worst hit. *Hindustan Times*. http://www.hindustantimes.com/india-news/india-proposes-ban-on-commercial-surrogacy-homosexuals-live-ins-worst-hit/story-Vb1fKz0XSJPdCT7GbympkO.html. Accessed 14 Aug 2017.

Harding, Sandra. 1991. *Whose science/whose knowledge?* Milton Keynes: Open University Press.

Hawley, Samantha, and Suzanne Smith. 2015. India surrogacy case: Documents show New South Wales couple abandoned baby boy despite warnings. *Australian Broad Casting*. http://www.abc.net.au/news/2015-04-13/australian-couple-abandon-baby-boy-in-india-surrogacy-case/6387206. Accessed 14 Aug 2017.

Henig, Robin Marantz. 2014. Women increasingly pick brains over looks in choosing egg donors. *National Public Radio*. http://www.npr.org/sections/health-shots/2014/10/30/360127832/women-increasingly-pick-brains-over-looks-in-choosing-egg-donors. Accessed 15 Aug 2017.

Hess, Amanda. 2014. The golden egg: Couples want their egg donors to be smart, athletic, and good-looking. *Slate Magazine*. http://www.slate.com/articles/technology/future_tense/2014/11/egg_donation_study_couples_want_donors_to_be_smart_athletic_good_looking.html. Accessed 15 Aug 2017.

Hindustan Times. 2017. Can't adopt white child, try in India, British-Sikh couple told by UK agency. *Hindustan Times*. http://www.hindustantimes.com/world-news/can-t-adopt-white-child-try-in-india-british-sikh-couple-told-by-uk-agency/story-d1q2x79j3psuvwntkupFrO.html. Accessed 15 Aug 2017.

Hubbard, R., and E. Wald. 1999. *Exploding the gene myth: How genetic information is produced and manipulated by scientists, physicians, employers, insurance companies, educators, and law enforcers*. Boston: Beacon Press.

Indian Council of Medical Research. 2015. National registry of ART clinic and banks in India. Indian Council of Medical Research: New Delhi http://www.icmr.nic.in/icmrnews/art/New%20list%20of%20approved%20ART%20Clinics_24.10.2015.pdf. Accessed 16 Aug 2017.

Inhorn, M.C. 2011. Globalization and gametes: Reproductive "tourism," Islamic bioethics, and Middle Eastern modernity. *Anthropology and Medicine* 18: 87–103.

Inhorn, M.C., and P. Patrizio. 2012. The global landscape of cross-border reproductive care: Twenty key findings for the new millennium. *Current Opinion in Obstetrics and Gynecology* 24 (3): 158–163.

Jadva, V., and S. Imrie. 2013. Children of surrogate mothers: Psychological well-being, family relationships and experiences of surrogacy. *Human Reproduction* 29: 90–96.

Kamin, Debra. 2015. Israel evacuates surrogate babies from Nepal but leaves the mothers behind. *Time*. http://time.com/3838319/israel-nepal-surrogates/. Accessed 14 Apr 2017.

Katz, Avi. 1986. Contract motherhood and the baby-selling laws. *Columbia Journal of Law and Social Problems* 20: 1–53.

Kirby, Jeffrey. 2014. Transnational gestational surrogacy: Does it have to be exploitative? *The American Journal of Bioethics* 14 (5): 24–32.

Knoll, E. 2012. Reproducing Hungarians: Reflections on fuzzy boundaries in reproductive border crossing. In *Reproductive technologies as global forum: Ethnologies of knowledge, practices, and transnational encounters*, ed. M. Knecht, et al., 255–282. Frankfurt: Campus Verlag.

Kozicka, Patricia. 2016. Reality check: Are fertility problems on the rise? *Global News*. http://globalnews.ca/news/3070201/reality-check-are-fertility-problems-on-the-rise/. Accessed 15 Aug 2017.

Krause, Elizabeth L., and Milena Marchesi. 2007. Fertility politics as "Social Viagra": Reproducing boundaries, social cohesion, and modernity in Italy. *American Anthropologist* 109 (2): 350–362.

Kroløkke, Charlotte Halmø. 2014. West is best: Affective assemblages and Spanish oocytes. *European Journal of Women's Studies* 21 (1): 57–71.

Kuchroo, Sakshi. 2016. Banning commercial surrogacy is the only way forward: Professor Mohan Rao of Jawaharlal Nehru University talks about the pros and cons of the surrogacy bill. *Governance Now* 7 (16): 38–39.

Kusum, S. 2015. Inconsistent and conflicting surrogacy laws in India and foreign legal jurisdictions. *International Journal of Legal Studies and Research* 72: 2278–4764.

Lippman, A. 1991. Prenatal genetic testing and screening: Constructing needs and reinforcing inequalities. *American Journal of Law and Medicine* 17: 15–50.

Mascarenhas, M.N., S.R. Flaxman, T. Boerma, S. Vanderpoel, and G.A. Stevens. 2012. National, regional, and global trends in infertility prevalence since 1990: A systematic analysis of 277 health surveys. *PLOS Medicine* 9 (12): e1001356. https://doi.org/10.1371/journal.pmed.1001356.

Maulem, Mazal. 2015. Israel airlifts surrogates, babies from Nepal. *Israel Pulse*. http://www.al-monitor.com/pulse/originals/2015/04/israel-nepal-rescuing-surrogate-mothers-gay-community.html. Accessed 15 Aug 2017.

Mohanty, Chandra Talpade. 2003. "Under western eyes" revisited: Feminist solidarity through anticapitalist struggles. *Signs* 28 (2): 499–535.

Murdoch, Lindsay. 2015. 'Somebody has to be the icebreaker': Aussies seeking babies turn to Cambodia. *The Sydney Morning Herald*. http://www.smh.com.au/world/somebody-has-to-be-the-icebreaker-aussies-seeking-babies-turn-to-cambodia-20151027-gkjfj5.html#ixzz3qMBjMidJ. Accessed 15 Aug 2017.

Nahman, Michal. 2008. Nodes of desire: Romanian egg sellers, 'dignity' and feminist alliances in transnational ova exchanges. *European Journal of Women's Studies* 15 (2): 65–82.
Nair, P.M., and Sankar Sen. 2004. *Trafficking in women and children in India*. New Delhi: Orient Longman.
Nelkin, Dorothy, and M.S. Lindee. 1995. *The DNA mystique: The gene as a cultural icon*. New York: Freeman.
Nelkin, Dorothy, and Laurene Tancredi. 1994. *Dangerous diagnostics: The social power of biological information*. Chicago: University of Chicago Press.
Nelson, Dean. 2012. Fair-skinned Indian women paid £1,000 extra to be surrogates. *The Telegraph*. http://www.telegraph.co.uk/news/worldnews/asia/india/9633142/Fair-skinned-Indian-women-paid-1000-extra-to-be-surrogates.html. Accessed 15 Aug 2017.
Nkrumah, K. 1966. *Neo-colonialism: The last stage of imperialism*. New York: International Publishers.
Nkwopara, Chidi. 2013. 14 pregnant teens rescued in another Imo baby factory. *Vanguard*. www.vanguardngr.com/2013/05/14-pregnant-teens-rescued-in-another-baby-factory-in-imo/. Accessed 15 Aug 2017.
Okin, Susan M. 1990. A critique of pregnancy contracts. *Politics and the Life Sciences* 8: 205–210.
Pande, A. 2009. Not an 'Angel', Not a 'Whore': surrogates as 'Dirty' workers in India. *Indian Journal of Gender Studies* 16 (2): 141–173.
Parajuli, Ramesh. 2015. Surrogacy in Nepal: Threat to reproductive right. *The Himalayan Times*. https://thehimalayantimes.com/opinion/surrogacy-in-nepal-threat-to-reproductive-right/. Accessed 15 Aug 2017.
Pateman, C. 1988. *The sexual contract*. Stanford: Stanford University Press.
Paul, Diane B. 1994. Eugenic anxieties, social realities, and political choices. In *Are genes us? The social consequences of the new genetics*, ed. Carl F. Cranor, 142–154. New Brunswick: Rutgers University Press.
Pennings, Guido. 2008. Ethical issues of infertility treatment in developing countries. *ESHRE* 1 (1): 15–20.
Pennings, G., C. Autin, W. Decleer, A. Delbaere, L. Delbeke, A. Delvigne, D. De Neubourg, P. Devroey, M. Dhont, T. D'Hooghe, S. Gordts, B. Lejeune, M. Nijs, P. Pauwels, B. Perrad, C. Pirard, and F. Vandekerckhove. 2009. Cross-border reproductive care in Belgium. *Human Reproduction* 24 (12): 3108–3118.
Perappadan, B.S. 2014. Activists call for stringent regulations for surrogacy. *The Hindu*. http://www.thehindu.com/todays-paper/tp-national/tp-otherstates/activists-call-for-stringent-regulations-for-surrogacy/article6348214.ece. Accessed 15 Aug 2017.
Pfeffer, N. 2011. Eggs-ploiting women: A critical feminist analysis of the different principles in transplant and fertility tourism. *Reproductive Biomedicine Online* 23 (5): 634–641.
Präg, Patrick and Melinda C. Mills. 2015. Assisted reproductive technology in Europe. Usage and regulation in the context of cross-border reproductive care. *Families and Societies Working Paper Series* (43): 43.
Prague Daily Monitor. 2015. Právo: No one wants Czech child born to surrogate mother. *Prague Daily Monitor*. http://praguemonitor.com/2015/08/13/pr%C3%A1vo-no-one-wants-czech-child-born-surrogate-mother. Accessed 14 Aug 2017.
Nagaraj, Anuradha. 2017. Wombs for rent: Indian surrogacy clinic confines women in 'terrible conditions', say police. *Reuters*. http://www.reuters.com/article/us-india-women-surrogacy-idUSKBN19A1KL. Accessed 14 Aug 2017.
Pujari, S., and Sayeed Unisa. 2014. Failing fatherhood: A study of childless men in rural Andhra Pradesh. *Sociological Bulletin* 63 (1): 21–40.
Qadeer, I. 2010. The ART of marketing babies. *Indian Journal of Medical Ethics* 7 (4): 209–215.
Rabinow, Paul. 1996. Artificiality and enlightenment: from sociobiology to biosociality. In *Essays on the anthropology of reason*, ed. Paul Rabinow, 91–111. Princeton: Princeton University Press.

Raymond, Janice G. 1993. *Women as wombs: Reproductive technologies and the battle over women's freedom*. San Francisco: Harper.

Riessman, C. 2000. Stigma and everyday resistance practices, childless women in South India. *Gender Society* 14: 111–135.

Rifkin, Jeremy. 1998. *The biotech century: Harnessing the gene and remaking the world*. New York: Penguin Putnam Inc.

Roberts, Dorothy E. 1996. Race and the new reproduction. *Hastings Law Journal* 47 (4): 935.

Roberts, Dorothy E. 2009. Race, gender, and genetic technologies: A new reproductive dystopia? *Signs* 34 (4): 783–804.

Robertson, John A. 1983. Procreative liberty and the control of conception, pregnancy, and childbirth. *Virginia Law Review* 69 (3): 405–464.

Robertson, John A. 1986. Embryos, families, and procreative liberty: The legal structure of the new reproduction. *Southern California Law Review* 59: 939–1041.

Rose, Nikolas. 2001. The politics of life itself. *Theory, Culture and Society* 18 (6): 1–30.

Rose, N., and C. Novas. 2005. *Biological citizenship*. Malden, MA: Global Assemblages, Blackwell.

Rothman, Barbara K. 1989. *Recreating motherhood: Ideology and technology in a patriarchal society*. New York: W.W. Norton.

Roy, Saurav. 2015. Trafficked tribal girls forced to conceive, deliver babies for sale. *Hindustan Times*. http://www.hindustantimes.com/india/trafficked-tribal-girls-forced-to-conceive-deliver-babies-for-sale/story-KEbZAB2au7Gyw9aspupTEI.html. Accessed 15 Aug 2017.

Sallam, H.N. 2008. Infertility in developing countries: Funding the project. *Human Reproduction ESHRE Monographs* 97–101.

SAMA. 2012. Birthing a market: A study on commercial surrogacy, New Delhi: SAMA Resource Group for Women and Children—unsure what this is/unsure how to cite.

Saravanan, Sheela. 2015. Global justice, capabilities approach and commercial surrogacy in India. *Medicine, Health Care and Philosophy* 18: 295–307.

Saravanan, Sheela. 2016. Humanitarian thresholds of the fundamental feminist ideologies: Evidence from surrogacy arrangements in India. Analize—Journal of Gender and Feminist Studies.

Sarojini, N.B., and Aastha Sharma. 2009. The draft ART (regulation) bill: In whose interest? *Indian Journal of Medical Ethics* 6 (1): 36–37.

Sartre, Jean-Paul. 2001. *Colonialism and neocolonialism*. New York: Routledge.

Shalev, Carmel. 1989. *Birth power: The case for surrogacy*. New Haven: Yale University. Press.

Sherlock, Ruth. 2016. Gay parents fight for custody with surrogate in Thailand. *The Telegraph*. http://www.telegraph.co.uk/news/2016/03/23/gay-parents-fight-for-custody-with-surrogate-in-thailand/. Accessed 15 Aug 2017.

Sherwell, Philip. 2015. India surrogacy ban dismays British couples. *The Telegraph*. http://www.telegraph.co.uk/news/worldnews/asia/india/12001903/India-surrogacy-ban-dismays-British-couples.html.

Shultz, Marjorie M. 1990. Reproductive technology and intention-based parenthood: An opportunity for gender neutrality. *Wisconsin Law Review volume?* 297–398.

Subedi, Madhusudan. 2015. Contractual transaction: How renting a uterus makes the human body a commodity in Nepal. *Dhaulagiri Journal of Sociology and Anthropology* 9: 1–25.

The Guardian. 2016. Surrogate mother who sold same babies twice sentenced for fraud. *The Guardian*. https://www.theguardian.com/world/2016/mar/23/surrogate-mother-who-sold-same-babies-twice-sentenced-for. Accessed 15 Aug 2017.

Times of India. 2010. Israeli gay case to hit surrogacy biz in India. *Times of India*. http://timesofindia.indiatimes.com/india/Israeli-gay-case-to-hit-surrogacy-biz-in-India/articleshow/5914884.cms. Accessed 15 Aug 2017.

Times of India. 2011. Womb for sale debates surfaces in Nepal. *Times of India*. http://timesofindia.indiatimes.com/world/south-asia/Womb-for-sale-debate-surfaces-in-Nepal/. Accessed 15 Aug 2017.

Times of India. 2012. Surrogate mother dies of complications. *Times of India*. http://timesofindia.indiatimes.com/city/ahmedabad/Surrogate-mother-dies-of-complications/articleshow/13181592.cms. Accessed 15 Aug 2017.

Times of India. 2017. Almost 20,000 women, children trafficked in India in 2016: Government report. *Times of India*. http://timesofindia.indiatimes.com/india/almost-20000-women-children-trafficked-in-india-in-2016-govt-report/articleshow/57569145.cms. Accessed 15 Aug 2017.

Vora, Priyanka. 2013. Indian couples seek 'white' donors for fair kids. *Hindustan Times*. http://www.hindustantimes.com/mumbai/indian-couples-seek-white-donors-for-fair-kids/story-ykVNvMJ7aPM4GPuGeBwpHL.html. Accessed 15 Aug 2017.

WHO. 2016. Sexual and reproductive health: Multiple definitions of infertility. *World Health Organization*. http://www.who.int/reproductivehealth/topics/infertility/multiple-definitions/en/. Accessed 15 Aug 2017.

Young, Robert J.C. 1995. *Colonial desire: Hybridity in theory, culture, and race*. London: Routledge.

Young, Robert J.C. 2003. *Postcolonialism: A very short introduction*. Oxford: Oxford University Press.

Zargar, Arshad R. 2016. India moves to ban surrogacy for potential foreign parents. *CBS News*. http://www.cbsnews.com/news/india-ban-surrogacy-potential-foreign-parents/. Accessed 15 Aug 2017.

Chapter 3
A Feminist Discourse on Surrogacy: Reproductive Rights and Justice Approach

Abstract Feminists have upheld the inclusion of reproductive rights such as using medical technologies for abortion, contraception and safe childbirth and these efforts need to be celebrated (ICPD Programme of Action in Cairo Egypt: international conference on population and development. United Nations Population Funds, 1994; UNFPA in Investing in people: national progress in implementing the ICPD programme of action 1994–2004. United Nation Population Fund New York, 2004). However, in the last two decades, developments in reproductive technologies and its use for practices such as sex selective abortions and pregnancy contracts has challenged some of the very ideologies that feminism fundamentally represents; equality (Structural inequalities and commodification in the surrogacy markets has been discussed in detail in Chaps. 4 and 5.) (socio-economic, health, legal), liberty (freedom of choice, autonomy) and justice (social and reproductive). Liberals supporting ARTs base their arguments around pro-choice, self-determination over one's body, liberty of using these technologies as women's empowerment, and contractual liberty (Inhorn and van Balen in Infertility around the globe: new thinking on childlessness, gender, and reproductive technologies. University of California Press, Berkeley, 2002; Zilberberg in Bioethics 21:517–519, 2007; Becker in The elusive embryo: how women and men approach new reproductive technologies. University of California Press, Berkeley, 2000; Roberts in Race and the new reproduction, 1996; Petchesky in Reprod Health Matters 3(6):152–161, 1995; Kishwar in Reprod Health Matters 1(2):113–115, 1993; Mies in Reprod Genet Eng 1(3):225–237, 1988). Feminists opposing reproductive technologies such as surrogacy are concerned that people, especially women, are becoming mere body parts in the flourishing global markets and that women may feel pressurized to become a part of it (Pfeffer in Reprod Biomed Online 23(5):634–641, 2011; Truong in ISS Working Paper Series/General Series 339:1–30, 2001; Gupta in New reproductive technologies, women's health and autonomy: freedom or dependency. Thousand Oaks, New Delhi, 2000; Shanley in Signs 18(3):618–639, 1993; Wichterich in The globalized woman: reports from a future of inequality. Spinifex Press, Melbourne, 2000; Roberts in Cyborg babies: from techno-sex to techno-tots.

S. Saravanan, *A Transnational Feminist View of Surrogacy Biomarkets in India*,
https://doi.org/10.1007/978-981-10-6869-0_3

Routledge, New York, 1998; Rothman in Creighton Law Rev 25:1599–1616, 1992; Corea in The mother machine: reproductive technologies from artificial insemination. Harper and Row, New York, 1985), causing commodification, exploitation, alienated labour, and denial of subjectivity (Hassan in WSQ: Women's Stud Q 38 (3):209–228, 2010; Dickenson in Property in the body: feminist perspectives. Cambridge University Press, Cambridge, 2007; Scheper-Hughes in Curr Anthropol 41(2):191–224, 2000; Sharp in Ann Rev Anthropol 1:287–328, 2000; Kimbrell in The human body shop the engineering and marketing of life, 1993; Swazey in Spare parts: organ replacement in American Society. Oxford University Press, Oxford, 1992; Raymond in Hastings Cent Rep 20(6):7–11, 1990). Scholars, activists, and some lawyers have questioned the individual reproductive rights approach in that it overlooks inequalities from a feminist social justice perspective (Donchin in Bioethics 24(7):323–332, 2010; Callahan and Roberts in A feminist social justice approach to reproduction-assisting technologies: a case study on the limits of liberal theory, 1996). This chapter reviews and critiques the liberal feminist approach of reproductive rights, drawing primarily on a reproductive justice framework.

Keywords Reproductive and contractual rights · Agency · Alienation
Altruistic surrogacy · Embodiment · Social stereotypes · Infertility
Stratified reproduction · Surrogacy as work

Liberal feminism draws heavily on the liberal political theory that focuses on rationality, individual autonomy, and choice (Shultz 1990; Robertson 1986). Reproductive right is defined as, "the basic right of all couples and individuals to decide freely and responsibly the number, spacing and timing of their children and to have the information and means to do so" (ICPD Programme of Action Paragraph 7.3). It includes access to contraception, abortion, and safe childbirth. The notion of individual autonomy is analogous to "my body my property" and comprises the option to freely use reproductive techniques as well as sell one's own body parts to third parties (Andrews 1986). Robertson (1983) defines procreative autonomy as "the notion that individuals have a right to choose and live out the kind of life that they find meaningful and fulfilling" (Robertson 1983: 230). Hence, Robertson (1983) justifies the use of technology for any reason that would realize the couple's reproductive goals. This is a utilitarian, or a realization-based, approach which emphasizes the ends rather than the means. Classic utilitarians, Bentham (1982) and Mill (1989), drew analogies between justified actions and pleasure. Utilitarians focus on actions that lead to general happiness (or less unhappiness). Although utilitarians express a desire for freedom, they are more interested in the consequences (general happiness) than the means to that happiness; the paradigm being women should be entitled to equal rights in employment opportunities, pay, and in political arenas (Williams 1982). Accordingly, some academics, activists, lawyers, and medical practitioners oppose the prohibition of surrogacy as it denies not only the liberty of intended parents but also the surrogate mother's right to enter into any contract that she may wish to enter (Katz 1986;

Andrews 1986; Shultz 1990; Robertson 1983, 1986). Raymond (1993) criticizes the procreative liberty defined by Robertson as institutionalizing commodification of women wherein women and children become objects and instruments to serve other's ends. In the context of India, where women consent to surrogacy due to poverty and are retained in dormitories for nine months, away from their families, there is no scope of choice; it becomes a compulsion. Shultz (1990) claims that developments in reproductive technologies, bringing personal intention and fulfillment in procreation and parenthood to the forefront, show that such advancements have some potential for further changes. She supports individual-based surrogacy so that this would recognize, encourage, and reinforce men's choices and yet contradicting herself she insists that intention-based reproduction is sex neutral. Possibilities of technologies, such as surrogacy, are assumed to enhance individual freedom, to control one's own destiny by embracing the ideology of autonomy. But ironically, surrogacy contracts require women to hand over the control of her body along with her reproductive experience and the baby to another person, reinforcing gender bias. Although socio-economic differences between intended parents and surrogate mothers are universal phenomenon, these differences can be more evident in developing countries (Harrison 2016; Blyth 1994). As long as individuals endure ARTs on their own body, the "my body, my decision" philosophy is relevant; however, when individuals begin to put another person through the same physical and psychological pain, and possibly even death, as is in the case of surrogacy, the philosophy of individuality becomes void.

ICPD defines reproductive rights as the right to life, privacy, and liberty, and individual rights to make informed decision about one's own body, to determine the number and spacing of their children, and be free from coercion, discrimination, and violence. This definition does not give reproductive rights to individuals to intrude into another person's reproductive function to fulfil one's own reproductive desire. Another clause is about individual's rights to non-discrimination and equality, discarding any bias of sex, race, ethnicity, language, religion, disability, and economic status. The third clause is on accountability of health, policies, and laws that include accountability of the functioning judicial system with the authority to subjugate sexual and reproductive rights violations. These reproductive rights are directed towards the responsibility of the state to safeguard citizens from state interventions that may violate human rights. According to these clauses, the state should not interfere with the right to autonomy, to eradicate discriminatory policies and practices and take affirmative measures to ensure that everyone is assured the same rights in law and in practice. It is assumed that only the state is capable of violating the basic reproductive rights of individuals, but what about individuals, clinics, or agents who violate human rights? What is missing is that it is not only the state but also individuals who should be held responsible and accountable towards non-discrimination and equality of all human beings. Individuals who discriminate other people's basic rights on the basis of sex, race, ethnicity, language, religion, disability, and economic status to fulfil their individual reproductive rights should be held responsible for the violation of human rights.

Individuals need reproductive rights but only with a sense of social responsibility and without violating other people's human rights. Individual rights to procreate do not give the right to violate other people's human rights and dignity. However, there is no clause in the ICPD to prevent individuals (intended parents, agents) and institutions (clinics, agencies) from being participants in the perpetration of discrimination and violation of human rights while fulfilling individual reproductive rights. CEDAW limits state interference with individual reproductive rights; but, what if such rights impede social justice and other's human rights? This reproductive liberty of the intended parents needs to be seen from the perspective of the agency of service providers and seekers in the ART market, surrogate mothers, gamete, and embryo donors. In the context of surrogacy, individualized rights of intended parents deprive another surrogate mother's right to equality, non-discrimination, dignity, and limit her decision-making regarding her own body, hence, violating another woman's basic human rights.

Instead, liberals urge for protection of the relationship between the intended parents and the child through the enforcement of surrogacy contracts (Shultz 1990). Robertson (1986) offers solutions towards safeguarding the intended parent's interest and emphasizing on the enforceability of the contract. The first solution accordingly is that the surrogate mother should be ordered to pay damages to the couple for any loss or suffering that the couple would have to endure if she aborted the baby; damages were also to be paid to the couple if she refused to relinquish the child at birth; further, as a trustee of the embryo for the couple she was also obligated to transfer custody at birth. This sounds like a typical courtroom judgement where everything is sorted out with compensation. Some feminists also have reiterated that women need to be held responsible for such contracts. Such a view does not take into account the socio-economic difference between the intended parents, the medical practitioners and the surrogate mother, the complexity of women's childbearing experience, and the possible inter-relatedness between the mother and the fetus. Surrogacy is not seen as a collaborative reproduction but rather a court battle where the surrogate mother is viewed as an employee whose terms of conditions are stipulated by the contract. She is not allowed to breach the contract, refuse medical intervention, abort the fetus against the wishes of the intended parents, negligently harm the fetus, and refuse to give up the baby. But the intended parents are granted full rights to force her into an abortion. Hence, surrogacy contracts are based on unequal market rules of belonging and ownership based on the genetic material. Some lawyers support intended parent's ownership over the baby even if the surrogate mother is genetically related to the child, such as the infamous Baby M Case from USA wherein the intended father was given custody although the surrogate mother refused to give away the baby.

Contractual Rights: Agency and Patriarchy

The notion of individual autonomy is analogous to "my body my property" and comprises the option to freely use reproductive techniques as well as the possibility to sell body parts to third parties (Andrews 1986). The rhetoric of choice "assumes that a woman's body is her own—that she owns it, controls it and makes her own decisions about her body, her health and her relationship" (Chrisler 2013). Katz (1986) supports traditional surrogacy because it guarantees a biological relationship of the child with the father, ability to choose a surrogate mother based on characteristics or phenotypes, and can provide a fixed waiting period to obtain the child. However, biological connection is not always sought, the ability to choose the surrogate mother'phenotype is a suggestion reinforcing racism, and the fixed period could be an illusion that there may be uncertainties as in any other pregnancy. Shalev (1989) contends that the surrogate mother is an autonomous agent and should be held responsible for the consequence of her autonomous decision to have a child for someone else. Hence, she should be also held responsible towards her contract like any other human being. Not being held to her contract implies her incompetence and inability to act as a rational agent by virtue of her biological sex (Shalev 1989). Shanley (1993) questions the certainty around women's voluntary informed consent. The self-identity of the body and mind is also shared as a social entity. In India, the social identity is more important than the individual self, wherein the body itself is construed as a shared entity. Reddy and Patel (2015) note that "my body, my property" also needs to be contextualized culturally especially when women enter into unjust surrogacy contracts because they have no other choice to earn money. Most women involved in surrogacy in India are in it for financial gains and these motivations are embedded in constraints (Rudrappa and Collins 2015; Pande 2010; Vora 2009; SAMA 2012; Saravanan 2015). This context is evident in the social reality that only 28.1% of married women in India take decisions on matters pertaining to the care of their own health (Senarath and Gunawardena 2009). Autonomy is considered essential for decision-making in a range of health care situations, beginning with seeking health care and utilization of available medical facilities till choosing treatment options. Among women in rural India, more than half (55.6%) were not involved in the decision-making regarding their own health care (Mistry et al. 2009). Women's autonomy in decision-making is positively associated with their age, employment, and the number of children living. Poor families are also more likely to keep girls at home to care for younger siblings or to work in family enterprises. Women in India may not assert their agency in basic human rights such as education, right to marry (when and whom), and health care facilities, especially women who have lesser educational background, and hence, in participating as surrogate mothers.

Informed Consent: India

In the context of surrogacy in India, the chairperson of National Commission for Women, Lalitha Kumaramangalam said that there is no informed consent in surrogacy; the only reason they (the women backed by her family) come into it is poverty, that is their only choice (NDTV 2015). India became a classic case of rampant violations of medical ethics practiced by clinics because surrogacy was unregulated within a permissive paradigm. Surrogacy black markets were thriving parallely, due to ineffective implementation. Commercial surrogacy became prohibited and restricted only to altruistic surrogacy since September 2015 in India. A question thereby arises: is an individual purely abstract, whose reasoning is "unaffected/uninfected by either the empirical-psychological ego or the empirical-biological body" (Jaggar and Struhl 1978; Scheman 1983)? Jaggar and Struhl (1978) and Scheman (1983) have thus criticized the liberal feminist notions as individualistic. Wendell (1987) supports individual rights by justifying that the abstract interpretation misunderstands the liberal feminist philosophy, maintaining that individual rights does not take an individualistic approach to morality and society. According to Wendell, a selfish person is one who takes more than his or her fair share of a resource. To fulfil one's individual reproductive liberty is one thing and individualism without consideration towards social responsibility is another. Carmel Shalev, an academician and feminist from Israel, in a documentary film, *Future Baby*, commented, "These (reproductive rights) were private decisions; hence state intervention in these was resisted. Having a right to be a parent is a positive right. But in recent days this is turning into an extreme version of consumer right, accessible to the rich. It goes rapidly from becoming 'a wish' to 'a desire' to 'a need' to 'a right' and 'an entitlement'" (Arlamovsky 2016).

Women, participating as surrogate mothers in India are coached into prioritizing the health of the child over her own health; they are made to sign the contract accordingly so that in the case of any untoward event, the child would be given priority. Women are also reminded repeatedly that the baby within them is not theirs but belongs to someone else that has to be given away. Women in India are known to prioritize their family over one's own needs. Gayatri, a surrogate mother in my study, on being asked if she would repeat surrogacy said, "Yes, I have gone through so much misery once (by participating in surrogacy) just for the sake of my children and I am willing to go through it again to save some extra money for my child's education" (Saravanan and Ranadive 2010). Her authority over her body may be an expression of **agency** but the nature and extent of that agency depends on her social circumstances. Similarly, in the context of sex selective abortions, women in Mira Nair's film *Children of Desired Sex*, spoke about repeated pressures to abort female fetuses said, "*by doing this only I will suffer, but others will be happy*" (Petchesky 1995: 402). She rightly questions: when women's identity and embodiment are so dissolved in kinship and patriarchal structures, how can the idea of "*my body is my own*" make rational sense (Petchesky 1995)? Although all women cannot be homogenized in one group, it is important to understand that women with such identities and embodiment cut across every class and region in India. Being an individual brought up in a middle-class Tamilian family in Mumbai, my body was not my own, I was free as long as my aspirations were limited to education and employment goals, while marriage, sexual freedom, and even reproductive rights was not included in this liberty but strictly bounded by familial and societal values.

As surrogate mothers, women in India experience a triple alienation binding them to different forms of institutionalized choicelessness; one form of alienation is poverty and inequalities that motivates them into surrogacy followed by a spatial alienation wherein they are supposed to be living in surrogate homes away from their family under restrictive conditions only breeding for the intended parents and having to think, eat, and behave in accordance to the requirements of the intended parents, while the baby grows inside them, and finally, having to alienate from the baby itself so as to maintain a distance from the baby growing inside (Hochschild 2011; Saravanan 2010; Vora 2009, 2010, 2013; Majumdar 2014). This dystopia of segmented reproduction is similar to the one Margaret Atwood described in her book *The Handmaid's Tale* and Gena Corea's division of reproductive labour (Corea 1985). In Atwood's book, no woman is whole, all individual women are reduced to divisive parts (Atwood 1985). These forms of alienation have an impact both physically and psychologically, not only on the surrogate mothers but also on their family and their children who are aware that their mother/wife/daughter-in-law is going through an extreme form of physically and psychologically traumatic experience for their wellbeing. They may be coerced into this practice by more powerful family members and may be forced to comply also in the hope of gaining more respect and love in the family which is nevertheless their right. Similar to the discourse on segmented reproduction, Scheper-Hughes (2000) observes that organ transplantation has been divided into two unequal populations—the organ receiver and givers. The givers are the anonymous, discredited supplier of spare parts, while the receivers are the cherished patients (comparable to intended parents) who are considered suffering moral subjects, retaining virtually unquestioned proprietary rights over the body parts of the poor, living or dead.

Glorification of Alienation

Some scholars have glorified the ability of women to be able to alienate themselves from the baby growing inside them as a liberating act that justifies that women are not emotional. However, a system that expects the surrogate mother to isolate herself from any emotion or attachment to the fetus growing inside her while considering the genetic connections as superior is a form of alienated labour, objectification, commodification, and denial of subjectivity (Saravanan 2010, 2013; SAMA 2012; Tieu 2009; Berkhout 2008; Van Niekerk and Van Zyl 1995). Requiring a surrogate mother to decide even before pregnancy about her parenthood expressions, during and after birth, represses any feelings that may possibly emerge towards the child during pregnancy or childbirth, and also giving others the power to hold her guilty if she diverges is alienation (Pateman 1988). As questioned by Anderson, "what if, despite her initial intentions, she finds herself coming to love her own child?" (Anderson 2000: 27). Hence, some of the arguments that emphasize on women's empowerment as reflected in the surrogate mother's ability and power to be able to detach from the feeling of motherhood while the fetus grows inside her, violates the integrity of women who may develop an attachment for the child. Scientifically, also there is evidence that the mother and child relationship cannot be completely alienated as it is now a known fact that cells migrate during pregnancy and this exchange occurs not only from the mother to the fetus but also from the fetus to the mother (Dawe et al. 2007). This means that all the babies and surrogate mothers have exchanged DNA (Deoxyribonucleic Acid) material during pregnancy. There is a debate among scientists whether or not DNA can be accepted as genetic material and if it is accepted then all surrogate mothers should be considered birth mothers and the entire definition of gestational surrogacy and descriptions of wombs as containers would then be questionable. Moreover, erasing a birth mother from her maternal identity and denying her such rights within surrogacy contracts can only be possible under the garb of patriarchy (Cornell 1998). The surrogate mothers in Amrita Pande's study in India felt a strong sense of blood relationship with the children they had borne and this was also revealed in my study where a mother commented: "the child is theirs but the blood is mine".

Several decades ago, Shultz (1990) supported individual autonomy on the basis of lack of any evidence of social harm in surrogacy practices. But now there is enough evidence emerging especially from transitional economies like India, Nepal, Thailand, Cambodia, and Mexico that surrogacy involves systematic exploitation of structural inequalities and violates human rights, women's body and integrity (Saravanan 2015, 2016; Whittaker 2011; Subedi 2015; DasGupta and DasGupta 2014; SAMA 2012; Rotabi et al. 2015). The practice of surrogacy in countries like India is embedded in structural and socio-economic inequalities. In surrogacy contracts, the intended parents own the gametes (self or donated) and the woman's body as well temporarily, and eventually, owns the baby she carries. Several writers have raised ethical questions of using a woman's social and economic vulnerability in India to commercially exploit her reproductive capacity. According to Dr. Nayna Patel "whenever somebody is rendering the service and the other person is ready to pay her the amount and when they are understanding adults entering into a contract, I don't see the term exploitation applicable over here" (NDTV 2015). Unfortunately, it is not merely exploitation but a sheer violation of human rights that is taking place within the surrogacy process in these clinics (Saravanan 2015).

Non-interference of State into Individual Privacy. Revisiting Altruism and Autonomy

The feminist position that supports women's access to and use of reproductive technology focuses on the personal experience of infertility (Sandelowski et al. 1990). The argument from this perspective is that the state should not have the right to interfere into a woman's will to participate in surrogacy. Liberal feminists have focused primarily on the reproductive goals of eliminating state-imposed gender distinctions and of preventing the state from limiting individual choice. These choices were defined to assist individuals with services and information for making reproductive decisions; however, these decisions are not merely personal or individual when someone else's life is also involved. Individual decisions to participate in surrogacy has a social impact and influences global reproductive justice by making use of transnational structural and gender inequalities. Others have noted that prohibiting surrogacy is also neocolonialism as it defies agency of egg donors and surrogate mothers (Fulfer 2017). By prohibiting surrogacy, scholars problematize domination over women's decisions, but by allowing surrogacy, domination is imposed not only over the surrogate mother but also on the unconceived child. It is assumed that the child will not yearn to know their genetic gestational roots. Dworkin (1983) notes that it is "the state (that) has constructed the social, economic, and political situation in which the sale of some sexual or reproductive capacity (becomes) necessary for the survival of women. The state denies women a host of other possibilities, from education to jobs to equal rights before the law". It is the state's intrusion into her selling of the body for sex or a sex-class-specific capacity that has provoked a defense for her will and her right; there should be more focus on providing women with the basic entitlements and human rights so that she doesn't have to sell her body in the first place (Dworkin 1983).

Some scholars support altruism while they are against commercial surrogacy (Kuchroo 2016). Andrew (1989) questions "why am I exploited if I am paid and not if I am not paid?". This argument presupposes that payments are restricted only to commercial surrogacy contracts. Dr. Nayna Patel clarified in a recent TV debate that "even (in altruistic surrogacy) within the family, payment is made in cash or kind. Someone bought a house, someone gave a car, someone gave a diamond set. Even in the altruistic surrogacy, it has never happened that intended parents did not do anything for the surrogate mother" (NDTV 2015). In a TV debate in India, Dr. Nayna Patel said, "nobody would be ready to surrogacy for someone else without money, among the 1120 babies born in my clinic through surrogacy, only 25 were within the family and they were not for free". Yet, the altruistic motive has been glorified by many scholars supporting surrogacy. If altruism was the prime motive for most surrogacy contracts, there would be a surplus of surrogate mothers in the UK and dearth of women willing to do this in India. Following a similar thought, Andrews (1989) supports payment for surrogacy because it is like any other service such as baby-sitting. Non-payment for surrogacy equates it with unpaid domestic work and banning payment for surrogacy would reinforce the state's power to define what constitutes legitimate and illegitimate reproduction, while payment would recognize a woman's legal authority to make decisions regarding the exercise of her reproductive capacity (Shalev 1989). Theoretically, this may sound like a logical argument but in practical terms it is inadequate in grasping the complexity of this practice. It has been observed, especially in countries like India, Thailand, Nepal that women who are desperately in need of money have their reproductive capacities used as a money-making machine by the clinics, agents, family members, and all those involved in the process regardless of whether the process is altruistic or commercial, catering to needs of the elite (refer Chap. 4). Some might argue by making these observations, researchers are slipping into the powerful/powerless dichotomy. But glorification of women's agency, when almost half of the women in India is not involved in decisions regarding their own body and have no access to health care, would lead us to what Amrita Banerjee described reductive fallacy (Banerjee 2011). The shortcoming of liberal arguments is to prioritize individuality over the symbolic body and its connectedness with human existence.

The Union Cabinet of India recently approved the Surrogacy (Regulation) Bill, 2016, according to which, commercial surrogacy is completely prohibited and foreigners cannot access surrogacy in India, but altruistic surrogacy is permitted only for needy married couples with the help of close relatives as surrogate mothers. This partial ban on surrogacy in India and also in Thailand has some grey areas; the glorification of altruistic surrogacy paves way for exploitation and the present selective ban discriminates against certain groups of people in society. It is known that altruistic surrogacy also involves money transfers and previous experience from other countries, like the UK, reveals that considerable amounts are transferred in the name of medical bills which can be exploitative to intended parents and surrogate mothers. Altruistic surrogacy can exploit women who may be dependent on other family members. Surrogacy between family members does not make the practice less exploitative than commercial surrogacy. Altruistic surrogacy within families, "romanticizes the family as the foremost place for protection" (Raymond 1993: 54). It is well known worldwide that most forms of abuse take place within close families and friend circles. It has been known that women in India tend to put other's needs and priorities before their own, which was evident among surrogate mothers in India who wanted to sacrifice their lives for the sake of the family (Saravanan 2013). With altruistic surrogacy, women within the extended family can be exploited to fulfil the reproductive needs of their affluent relatives, while women who refuse may face ostracism. Radical-socialist feminists have critiqued altruistic surrogacy as a compassion trap in which infertility is portrayed as a desperate need (Tong 2009) and an appeal is made to generous, loving, altruistic women to step forward to give the gift of love, a bundle of joy, to sorrowing, lonely, and childless couples. However, the same loving woman who is expected to feel altruism towards the intended couples should very generously sever all ties with the child after birth for the rest of her life. Accordingly, in India, women who are submissive are chosen while assertive women are refused (Saravanan 2015). It is observed women who choose to be a surrogate mother are those who are guilty conscious of having aborted or given their child away for adoption or want their self-esteem to be enhanced by this humanitarian act. Both of these motivations need other kinds of support such as psychological care rather than surrogacy that may cause more psychological damage.

Marxist and radical feminists criticize that the lower socio-economic status of surrogate mothers and prostitutes drives women to make themselves biologically available as that is their most valuable possession (Tong 2009). In her choices between two evils—being poor and being exploited, she may choose one evil so that she is able to overcome the other evil. Although class differences are more visible in India, surrogate mothers all over the world, even for altruistic surrogacy, are more likely to be selected if she is poorer and in need of money. And, when women are willing to do the same for one-third of the cost in less developed countries, why would there not be a flourishing market for it in such countries? Radical feminists additionally say that women are socialized to meet reproductive and sexual desires as a matter of duty and pride. Women in India take on surrogacy as the only choice to remove their families from the vicious cycle of poverty.

Indeed, there is a surplus of women wanting to be surrogate mothers in countries like India. The embodiment of value and reward that women in India experience as a form of agency within the surrogacy process will be discussed in detail in Chaps. 4 and 5. A large proportion of people do not have access to essential basic needs such as food, energy, housing, drinking water, sanitation, health care, education, and social security so that they are able to achieve a decent standard of living. In this situation of bare subsistence, their choice is between poverty or surrogacy. A similar surplus of surrogate mothers is not observed in affluent countries like the UK allowing altruistic surrogacy. Protest needs to be directed towards enhancing the essential needs of people in transitional economies like India. Any activity that violates a person's dignity or integrity and involves economic exploitation would not be considered a constitutional right (Raymond 1993). She observes that viewing reproductive technologies and contracts as a woman's choice emerges from a Western ideology of individual freedom. Western individualism entailed in liberal theory drives liberal feminism relying on a universalistic morality that misconceives contextual structural injustices manifest in poverty and deprivation. However, over-contextualization also leads to extreme polarization and there needs to be a balance between zooming in and out of situations.

Robertson (1986) fears that laws satisfying the offspring's needs to know their roots might prevent the birth of future children through collaborative arrangements such as surrogacy. He says a deracinated existence would be preferred to not being born at all. He also considers the privacy rights of the donor or surrogate be respected over the child's. This is precisely why Radin (1995) suggested baby-selling and surrogacy as two of the three forms of labour that should not be traded in markets and coined the term "market-inalienability".

The concept of the individual is an isolated entity, while surrogacy contracts are embedded in social relations and embodiment that are non-contractable (Held 1987). Scheper-Hughes and Lock's (1987) mindful body deconstructs embodiment into three perspectives: as phenomenally experienced by the individual body-self; as a social body, a symbol of social/structural relationships between nature, society, and culture; and as a body politic, an artefact of social and political control. Scheper-Hughes' more recent work focuses on the trafficking of human organs in global markets, health inequalities, embodiment of social forces and the commodification of human bodies (Scheper-Hughes 2000). Young's (2005) pregnant embodiment examines embodied experience of; infertility, pregnant person, relationship between the self and the embryo-fetus-forthcoming child. Douglas's (1970) body is a symbol to understand the social interaction between the self and the society; Rose's (2001) biopolitics refers to risk politics with a focus on biological and genetic traits determining human (in)capacities. Hence, the "my body my own" concept is inadequate in understanding the body in its entire complexity.

Katz (1986) describes embodiment as a free will but limited to the extent of signing the surrogacy contract. Some scholars emphasize on embodiment between the surrogate mother and the intended mother as sisterly, the surrogate mother with the fetus merely as incubatory, with less mention about the embodiment with the child born (Teman 2010; Pande 2014). But embodiment has many layers and is far more complex than the act of signing of the contract. Those who support altruistic surrogacy base their argument on the limited body–mind understanding of embodiment. There are also several other forms of embodiment observed. Most studies in India revealed that surrogate mothers felt a motherly embodiment towards the baby in their womb, and also as a connection of blood (Pande 2014). Additionally, in my study, a few surrogate mothers told me that they have observed other surrogate mothers who experience an aversion towards the entire idea of surrogacy, as being a financial compulsion (see Chap. 5); this may be psychologically detrimental for the surrogate mother and the fetus. The clinic insists that surrogate mothers spend their time in the surrogate home reading religious texts which according to them keeps their mind occupied in the principles of good karma and in doing so impose another level of rules in an attempt to control their minds.

Infertility is known to engender self-esteem/identity alongside feelings of hope, anticipation, worthlessness, (dis)satisfaction with life, anger-resentment, grief-depression, anxiety-stress, envy-isolation, a lack of personal control, and a loss of the dream of co-creating, especially among women, which can persist for as long as 20 years after discontinuing treatment (Wirtberg et al. 2007). Social stigma around infertility is a universal phenomenon; in industrialized societies, it is experienced as silenced stigma, while in other cultures it may not be so hidden (Greil et al. 2010). In extreme cases, infertility can lead to marital failure, physical or emotional abuse, social exile, and poverty (Inhorn 2009). Those experiencing some of these emotions, especially women, often are known to obsessively seek treatments to remedy the state of childlessness to fulfill a cultural norm (Whiteford and Gonzalez 1995; Schroeder 1988). Women tend to view infertility as a greater tragedy than their male partners and feel more stigmatized, tend to carry a larger physical burden as their bodies are subjected to invasive infertility treatment even when the pathological symptom is experienced by men (Whiteford and Gonzalez 1995; Sandelowski 1991). Thus, although male infertility accounts for more than half of the cases worldwide, majority of the research focuses on women as it primarily tends to be socially constructed as a women's burden. Women were more likely to use words such as "failure" and "broken" to describe their bodies, while men have referred to accepting the failed treatment but relating their experience to emasculation. As infertility becomes socially construed as a medical condition, an illness, even a form of disability, there is an increased pressure on people, especially women to avail to these reproductive technologies and they become consumers of this technology wherein it is difficult to spot their agency (Ginsburg and Rapp 1995).

Embodiment has several layers and is all-encompassing. The individual body is understood in the phenomenological sense of the lived experience of the self which assumes that all humans experience at least some intuitive sense of the embodied "self" (Mauss 1985). Young's pregnant embodiment allows the conceptualization of the body-self in experiencing infertility, childlessness and pregnancy using phenomenology of the body that interacts with the social (Young 2005). There is an ongoing debate between those who are concerned about the commodification of the body and body parts and those giving precedence to free market participation and altruistic element in the PGD (pre-implantation diagnosis) selection practice (Krones and Richter 2004). According to Appadurai (1986), things (biomaterial) turn into a commodity at a certain phase when they are exchanged for an economic value, which is the value assigned to things by subjects. The second, social body is the connection between an individual's physical body with the social world of the individual. This level of analysis refers to the representational use of the body as a natural symbol with which society and culture is perceived as suggested by Douglas (1970). Scheper-Hughes and Lock's (1987) idea of social body follows the path of social, symbolic, and structural anthropologist thinking of exchange of the meanings between the natural and the social worlds reinforcing each other. In analyzing the elusive body and carnal sociology, Merleau-Ponty (1995) contends that both *what is done to the body* and the active role of *what the body does* should be taken into consideration to understand the social habitus of the body-subject. For Foucault, the body that acts is important while Merleau-Ponty emphasizes on the acted-upon body. Crucial to Young's existential phenomenological understanding of the body in pregnant embodiment is the understanding that embodiment is a mode of "*being-in-the-world*" wherein bodies are impacted by social norms that govern them (Young 2005, 9). It is power that causes individuals to police their own bodies to ensure a maintenance of social norms that impacts one's lived body (Bordo 1993). Alcoff clarifies within phenomenological accounts that there is a necessity to consider the impact of cultural and ethnic specificity (Alcoff 2006: 107). Lamont et al. (2015) defines symbolic boundaries as practices that include certain people and groups while excluding others through patterns of likes and dislikes, cultural attitudes and practices. The third level is the politic body refers to the regulation, surveillance, and control of bodies (both individual and collective) in reproduction and sexuality. Biopolitics is a complex concept, as developed by Foucault, that examines the strategies through which human life processes are managed by regimes of authority (Foucault 1997). Rose (2001) refers to biopolitics as risk politics, raising concerns on the resurgence of biological and genetic traits determining human (in)capacities. Towards managing risk politics, authority acts upon improving the body politic by relieving the social and economic burdens of disease and disabilities (Rose 2001).

Women in India participate in surrogacy to enhance their standard of living; medical practitioners and other intermittent agents are in this business mainly for profit making; and intended parents from abroad come to India because of the restriction in the source countries, the prevailing lower costs and lesser rights for surrogate mothers in India. However, surrogacy in India is embedded in the complex social context of structural inequalities, gender roles, notions of family, and perspectives of the body in terms of the self and the socio-politic body. In India, women's understanding of their body in terms of the 'self' is very limited as compared to the socio-politic requirements of their body. Women are generally expected to put their body through arduous torture to assist their family live a comfortable life and in return receive affection and respect from their husband and in-laws. The woman's reproductive capacity is comparatively more remunerable than men. Moreover, if women had better education, they would have had better job opportunities and may not have had to involve in surrogacy practice. In comparison, in most Western countries basic education and health care is provided to every citizen as a basic human right. Hence, the relationship between family members is embedded in unequal gendered roles which women (with fertility capacity) strive to enhance by involving in surrogacy. The contextual understanding of family is deeply embedded in the cultural notion of motherhood (surrogate and social). In ancient Indian literature and in contemporary times, both the social mother and the surrogate mother are considered significant persons in the upbringing of the child as evident in the bringing of Krishna and Balaram, the most widely revered and popular Indian divinities in Hinduism.

Hence, in the Indian cultural context it is known that a child needs to have the blessing of both these mothers to lead an accomplished and contended life and the child needs to be obliged to both these mothers. The findings of my study reveal that the surrogate mothers wanted to keep contact with the child(ren) and to know about their well-being and progress. Most intended parents however were advised by the medical practitioners in India to sever all contacts with the surrogate mothers. Some surrogate mothers experience a soul connection with these children. For example, one surrogate mother related her body experience of hair loss with growth (teething) of the child(ren). Many other celebrate the birthdays of these children and feel the pang of separation for several years after relinquishment. Frustrated with the severed contact with the intended parents and the children, some surrogate mothers in my study even called the intended parents "*ehasaan faramosh*" (ungrateful). Others mentioned that, given a choice, they would prefer that their children are sent back to them. All surrogate mothers I interviewed in India preferred an open surrogacy in which they would be allowed to meet and interact with the intended parents and with the children born. However, the medical practitioners present surrogacy as an unattached "work' by the surrogate mothers and convince the intended parents to sever all contacts with the surrogate mothers after payment. Accordingly, some intended parents do not even meet the surrogate mothers face-to-face even for the payment. The medical practitioners in India perceive the practice from a business point of view. They portray remuneration is a primary motivation for the surrogate mothers and underplay the bonding with the children that is context-specific. The informed consent of the surrogate mothers is questionable given the gendered embodiment and the structural inequalities between the intended parents, the medical practitioners, and the surrogate mothers. It is important to understand these complex notions of families and inequalities in transnational commercial surrogacy practice in India.

The new surrogacy Bill (2015) limits surrogacy only to married heterosexual couples who have tried for children for four years. This selective prohibition of surrogacy to certain groups of people (homosexuals, single parents, live-in couples) reinforces patriarchal and homophobic stereotypes. Mohan Rao (in Kuchroo 2016) critiqued the Surrogacy Bill in India suggesting that altruistic surrogacy "should be allowed for everybody whether they are lesbians or gays is irrelevant—and to live-in couples." However, in this globalized world, it is capitalism and structural inequalities that play a primary hegemonic role in commodifying human bodies.

It is only the affluent people from various sections of the society who can afford surrogacy. The socioeconomically disadvantaged people in the society regardless of whether they are married, single, infertile, same-sex couples or eunuchs cannot have children through surrogacy whether it is India or any other country in the world. The sympathy towards homosexual and single couples seems to takes precedence over the violation of human dignity and human rights.

Reproductive Liberty, Patriarchy, and Infertility

Intended parents are known to approach assisted reproductive technologies to fulfil their childbearing desires that are embedded in social and self-identity of infertility[1] (Letherby 2002; Franklin and Roberts 2006). Selection of surrogate mothers and embryos and gametes within the surrogacy process is made after thoughtful and thorough personal and social considerations (Ehrich and Williams 2010). Increasing availability of technologies, however, has put a pressure on individuals, especially women, to try these choices. In the process, women become consumers of this market; "it is hard to spot the agency of women in the development of a technology for which they become consumers" (Rapp 1999: 34). Moreover, infertility itself is posed as an individual experience; hence, social solution to the problem is not sought (Sandelowski et al. 1990).

Male infertility accounts for more than half of the cases worldwide, however infertility is still seen as a woman's social burden (Inhorn and Patrizio 2012: 1). This is due to the focus on the female reproductive system and construction of the woman as the infertile patient (Carmeli and Birenbaum-Carmeli 1994: 666). However, infertility affects men as well and can be equally as devastating even though the responses and reactions are altered through the differential gender roles, expectations and socializations (Whiteford and Gonzalez 1995; Inhorn 2002). Now that ARTs are available, the cultural understanding of infertility has shifted and couples are expected to take advantage of these medical interventions (Whiteford and Gonzalez 1995: 27).

More is written and spoken about women's infertility than men; women also tend to carry the burden the infertility more than men both in terms of social pressures and psychological impacts (Nahar and Richters 2011). Men have expressed a lack of support and ambivalence of social support in their infertility experience requiring a holistic approach to infertility issues including expressions of male engagement with their reproductive bodies (Schick et al. 2016; Gutmann 2007; Sloan et al. 2010). While some feminists relate the concentration of infertility treatment and academic discourse around women's experience of infertility to male

[1]Infertility is "a disease of the reproductive system defined by the failure to achieve a clinical pregnancy after 12 months or more of regular unprotected sexual intercourse." (Zegers-Hochschild for WHO-ICMART 2009).

dominance, according to Carmeli and Birenbaum-Carmeli (1994), this is because men find themselves at a disadvantaged position in infertility treatments because of the male-inspired association of procreation with women, the perceived threat of male dispensability due to availability of sperm donation and scarce medical treatment offered to men. However, it is women who bear the burden of their own and the male partner's infertility not only socially but also in the medical interventions of IVF.

Societal pressures to use these technologies are also embedded in this identity and notion of infertility. Women tend to feel desperate and unfulfilled owing to their infertility experience; however, this sense of lack is given to them by the social pressures (Saravanan 2017). Some studies relate infertility to malnutrition in developing countries and to postponement of motherhood in developed countries. Many writers have related the social stigma towards infertility to be a more serious problem in developing countries than in industrialized societies (Van Balen and Bos 2009). Others observe that social construction of infertility is a universal issue although in industrialized societies infertility is a "silenced stigma", in other cultures it may not be so hidden (Greil et al. 2011a, b; Johnson and Fledderjohann 2012). For instance, in India there is a strong preference for heteronormativity. The typical nuclear family focuses on the relationship between the man and woman, and it is expected to be a reproductive relationship resulting in children (Bharadwaj 2016). When infertility occurs, these identities and family structures are questioned often resulting in stigmatization due to pressure to conform. Infertility and childlessness are considered unnatural and lead to "an 'unfulfilled' life for the woman [and] 'emasculation' for the man" (Bharadwaj 2016). Infertility is hard on both men and women; however, often childless women suffer more as they are a focus for patriarchal power resulting in social repercussions (Widge 2002). Indian women are constrained by pronatalism and the notion of compulsory motherhood, they are aware that most of their power and agency is derived from their reproductive capacities (Riessman 2000; Pujari and Unisa 2014; Dube 1998; Sayeed 1999). Thus, women in Indian society are defined by their fertility and can feel worthless when faced with involuntary childlessness (Widge 2002). Due to these vast social pressures and internalized patriarchal norms, women (and men) will do all they can to conceive a child of their own (Widge 2002).

Social Stereotypes of Motherhood and Infertility

However, a comparatively developed country such as Germany is known to be a strong pronatalist country, and women (and couples) face stigma if they do not reproduce (Greil 1991; Miall 1986; Riessman 2000). Within Germany there is a trend towards postponing childbirth or parenthood which stems from a variety of reasons, including advances in reproductive medicine and the belief of the ability to have a child at any time with assistance (Brähler and Stöbel-Richter 2002; Stöbel-Richter et al. 2005). However, alongside this trend is that of **infertility**, which evokes similar reactions compared to India, stigmatization and secrecy (Mishra and Dubey 2014). Infertility in Germany is considered, generally, as the woman's issue and that "men delay accepting the need for treatment for infertility," which is similar to how men in India deal with infertility (Unisa 1999: 56). The use of ARTs to conceive via third party are slowly becoming more normalized in Germany, however there is still a lot of privacy and secrecy surrounding it (Wischmann and Thorn 2013). As there is an urge to identify infertility as a form of disability and there is increasing possibilities being made available through IVF developments, the pressure on people to avail these reproductive technologies will increase and so will the social stigma of infertility (Khetarpal and Singh 2012; Ginsburg and Rapp 1995).

Some academics consider surrogacy contracts as defying common analogies drawn between femininity and nature and the definitions motherhood by being able to define their own pregnancy in their own terms (Teman 2009; Shalev 1989); others consider surrogacy as an arrangement that reinforces stereotypical notions of motherhood and women's social roles (Berend 2010; Roberts 1998). Feminists have strived for women's freedom from their stereotypical motherhood role in the society to be able to participate in the public sphere to follow non-reproductive aspirations. The surrogate mother's ability to separate herself from the fetus is described as a liberating experience that increases her autonomy (Baker 1996; Teman 2009; Shalev 1989). But liberation from such stereotypical roles proved by participating precisely in the same roles (reproduction—paid or unpaid) for someone else is highly questionable. As women have gained entry into greater socially valued forms of work, the severity of childlessness seemed to diminish (Donchin 1996). However, with access to IVF treatment women face different kinds of pressures; infertile women face more pressure to produce biological children while other women with reproductive capacities in the developing countries like India face the pressure of rescuing their families out of poverty (Donchin 1996; Saravanan 2016). These women have been absorbed into this already devalued reproductive domain as service providers and face various violations of bodily integrity such as signing off rights over the baby even before they enter into the contract, multiple embryo implantations, selective in utero abortions, compulsory caesarean section, compulsion to stay away from their family, and so on.

Surrogacy also reinforces stereotypical views that perceive infertility or childlessness as a social stigma. Most people seeking surrogacy desire children not merely for themselves but also because they have endured social stigma due to childlessness and now want to regain social acceptance and dignity. It is the social context of pronatalist sentiments that generates the need for infertility to be considered a disease and recently also a disability and infertile persons are urged to fulfil their full reproductive potential using ARTs which are generally directed at women regardless of the economic, psychological, and bodily costs. In India, it is known that having a baby is valuable for women as they are symbols of motherhood and it increases their honour and esteem in the eyes of the in-laws, especially so on the birth of a boy child (Mishra and Dubey 2014). This is reflected also in surrogacy practices in India, as it has been observed that having a boy child would entail higher bonus amount for medical institutions, brokers and the surrogate mother as intended parents tend to be more contended. Surrogate mothers in India are also blamed in case the child has some problems (disability or sometimes even for the birth of a girl child). It is understood that they would not be paid any bonus or sometimes they are not even paid the full amount promised in the contract with the birth of a child with disability.

Another stereotypical role of women that surrogacy reinforces is the tendency to put other's need and priorities before their own (Baker 1996). This motivation is evident among surrogate mothers in India who want to sacrifice their lives for the sake of the family (Saravanan 2013; Pande 2010). The surrogacy markets hence operate amongst those who (consciously or unconsciously) subscribe to patriarchy (a subordinate position of women in society) and they may not always necessarily be men. These may include lawmakers, medical institutions, surrogate mothers themselves, intended parents, and brokers consisting of people from different gender categories including women. Another instance of state-enforced patriarchy is the example of Israel, where the surrogacy laws is strongly patriarchal (Shalev 1998). Patriarchal control and racial privilege is reinforced over women's reproductive bodies through law implementation. Only people who are married, heterosexual and both Israeli partners with a Jewish descendent proof, can opt for surrogacy in Israel. In Israel, couples have to go through psychological and genetic testing to prove that they are not "unsuitable" so as to prevent the birth of "unhealthy children" (Shalev 1998).

Geneticization

The owner of the body holds the right to prevent harm as well as to earn profits from one's own body parts. Dr. Nisha says, "it is not womb for sale, we tell women that they are renting their body to hold someone else's baby for some time to be handed over at the end of nine months." She is partly right because this is not comparable to organ donation, surrogate mothers are not donating/selling their womb but carrying a baby for another couple. A child is developed during this time; selling a body part is one thing and a baby growing inside can give rise to several reactions. The surrogate mother is expected to give away this baby at the end of the contract and in India she has to sign off her rights/emotions towards this even before she enters into the contract. One concern is about expecting all women to decide before their pregnancy about the sort of relationship they would develop during the pregnancy. The second concern is about the tendency towards handing over the legal custody of the child to the intended parents even if the surrogate mothers ask for custody after the birth of the child. From the perspective of some feminists, a prohibition on surrogacy would imply that women are incompetent by virtue of their biological sex to make rational decisions regarding their reproductive activities (Shalev 1989). She considers those who allow surrogate mothers to change their mind fall into the age-old trap of women being irrational and can be driven by sentiment. “Contract is a contract, we (women) are not ruled by our hormones” suggests Andrews (1989). Others insist that women should not receive any protection unavailable to men except maternity leave (Finley 1989; Littleton 1987).

Conversely, Anderson (1990) notes that if women are expected to repress from establishing any emotional attachment with the child, it is a form of alienated labour. In terms of market norms, it may be legitimate but when these norms are applied to women’s reproductive labour, women are reduced to objects of use (Anderson 1990). Can we say that all women involved in surrogacy alienate themselves emotionally and physically from the baby she carries during the surrogacy contract? Some do and some don’t and the numerous cases that the surrogate mothers have filed custody for the children is an evidence of their bonding and a feeling for the well-being of the child(ren). However, it is most often seen that after the child is born, geneticization plays an important role in prioritizing and legally handing over parenthood custody to the intended parents as the rightful parents of the child(ren), while downplaying the gestational role of the surrogate mothers. Hill (1990) supports for surrogacy arguing that it was the intended parent’s interest to commission the contract that brought the arrangement into existence in the first place, hence, priority should be given to them in custody cases.

The case of India is significant in proving that if freely permitted, this baby market can stoop to very low standards. In India, in the clinic that I visited and several other clinics, it is mandatory that the surrogate mother signs off all her rights over the child in the contract. In my study, all clinical records, from the registration of pregnancy, health records, medical bills and, ultimately, to the birth certificate were registered as a pseudonym or in the name of the intended mother without any mention of the role of the surrogate mother. Her name was mentioned in the contract and a copy of that contract was not given to her. With no knowledge regarding her (very limited) rights in the ART Bill, no legal support whatsoever and with no financial support to fight any case in the court of law, surrogate mothers in India don’t want to follow that path.

Analogy Between Surrogacy and Prostitution: Normalizing Both as Work

The analogies drawn between surrogacy and prostitution is not about morality but the similarity of the discourse. Marxist feminists observe that surrogate mothers just like prostitutes are of lower income categories in comparison to their clients that produces alienation through the capitalist system. Mohanty (2003) draws attention to the continuities and discontinuities between the haves and have not both within and between national boundaries. Intended parents seek surrogacy and gametes in countries such as India not merely for regulatory inconveniences in the home country but also for cheaper supply of gametes and so-called gestation carriers and the control they enjoy over the surrogacy. This is not to say that the surrogate mothers in India do not engage in any sort of agency but the extent of their agency is highly questionable. Driven by poverty, women are compelled to sell the only valuable thing she owns, her body. Surrogate mothers, in my study, were involved in drug trials[2] for money another means of selling one's body. Barbara Katz Rothman observes that like prostitution surrogacy is an intimate relationship but the difference is that the relationship is mainly with the baby and this relationship according to her is not surrogate; it's a lived experience. In both these activities, women's bodies are more remunerable; hence, women from poor socio-economic background become more easily bioavailable to this market. The other similarity based on this is that most women who choose these activities (prostitution and surrogacy) as an earning option do so because of poverty and as it pays better than any other work available to young, unskilled and often illiterate women (Hochschild 2004). Hence, poverty assumes a subtle form of control that compel them to sell their body (MacKinnon 2011). The capitalistic and commodification patterns make certain forms of bodies more available, a hierarchy that determines that mainly developing countries contribute to the supply side and the developed countries participate in the demand side. People turn themselves into commodities that broadly eliminate their ability to influence their conditions transforming them into a commodity that forfeits the power to control the terms of exchange. Women face severe forms of human rights violation in both these activities. Some scholars have argued that surrogate mothers described their pregnancy as an enjoying experience, but this was not observed in the Indian context, rather surrogate mothers in India were saying that they were doing it out of desperate compulsion of money and they would never want their daughters to do this.

Most of the women involved in prostitution are trafficked and most women willingly participate in surrogacy knowing that it is exploitative in order to enhance the socio-economic status of their family. Hazel Thompson, a British photo-journalist who spent a decade documenting the lives of girls trafficked into India's sex industry calls it "modern day slavery" (Thompson 2013). One of her main observations is that majority of the women in Kamathipura, the popular prostitution area in Mumbai, are trafficked. Both these industries fall under the informal sector in India. In recent days, technology is forcing a change and so are regulations such as demonetization due to which the informal sector is made to use electronic transfer of money in India. It's not just the technology but the policy changes in India that compels people to pay monthly salaries to various household helps (housemaids, cooks, nannies, plumbers, electricians, painters) through direct bank transfers (Padmanabhan 2016). Although this is only the first step towards formalizing the informal sector, the beggars and prostitutes are at the bottom end of

---

[2]The number of approved Global Clinical Trials (GCTs) in India rose sharply from 65 in 2008 to 391 in 2009, which continued to rise with 500 GCTs being allowed in 2010, 325 in 2011 and 262 in 2012. Recently, the revelation that 2,262 people had died in these trials during the past five years led to a public outcry and the Supreme Court intervened with stricter norms for controlling drug trials. Indians were being used as guinea pigs in these drug trials and the Supreme Court had criticized the Health Ministry for allowing this.

the hierarchy. The present Indian government has made several policy strides in this direction; Jan Dhan Yojana and Aadhar card urging people to open bank accounts and the identity card that has the possibility to catch/trace trafficking (Philip 2017). Bank account and digital transactions for salaries can be one small but definite way forward towards formalizing the informal sector.

With the legalization discourse for and against surrogacy, is the tendency to normalize violations of human rights and that surrogacy practice will follow the failure of the law implementation evident with the prostitution regulation in India. According to estimates, the large majority of India's estimated 1.2 million prostitutes are forced into the trade by abject poverty. According to the regulations related to prostitution in India 1956,[3] prostitution per se is not illegal or prostitutes are not criminalized but third party facilitation of prostitution such as brothel keeping, living off earnings, and procuring is punishable. It is estimated that there are 275,000 brothels operating in India. According to the immoral traffic (prevention) Act, 1956, higher punishment is stipulated for inducing prostitution where the offence involves children. The number of prostitutes is estimated to be 2 million and children constitute half a million of these. In six decades, this market-driven industry has not been regulated enough to reduce child trafficking, crime, and violence in this sector. The other factor is the unregulated informal sector in India, under the purview of which labour using women's body such as prostitution and surrogacy fall.

The informal sector in India accounts for approximately 70% of the labour force (ILO 2012). Others estimate this sector to consist of 90% of India's workers, meaning a vast majority of employed people do not have the privileges of social security and workplace benefits. In the list of informal employment sector in India, prostitutes are not even listed among earners, rather they are listed along with beggars among non-market free employment category (NSSO 2012). The immoral traffic act has been an abject failure. India, as a country is not mature enough to handle legalization of such sensitive forms of exploitative women's labour and until this standard has achieved legalization, surrogacy can be a harrowing experience with unregulated slavery-like conditions within the contract. Regardless of whether it is legalized, nobody cares about the labour conditions of women's sex and reproductive labour. Concerns have been expressed that banning surrogacy would drive the practice underground. All these years, the prostitution industry in India is based on trafficking black markets, and despite being partly liberalized, nobody is concerned. Even in countries with a comparatively better regulatory body in place such as Germany and Holland, legalising prostitution has led to illegal market that was four to five times the size of the legal market in 2007 (Klein 2017).

[3]Laws related to prostitution in India: Suppression of Immoral Traffic in Women and Girl Act-1956, Prevention of Immoral Traffic Act-1956, Immoral Traffic (Prevention) Act-1956.

If India enhances regulation, labour rights, and salaries for workers in the informal sector such as; domestic work, garment factories, there will be fewer women seeking options such as surrogacy, which is exposing women to extreme forms of human rights violation and mortality risks. It is important to understand that markets such as surrogacy mushroomed rapidly, especially in places where slavery-like employment sectors, child sale and trafficking is already rampant and unregulated. It is estimated that about 10,000 children are trafficked from Jharkhand state in India every year to work as domestic help or sex workers, and trafficking for forced surrogacy is a new entry into this market based on women's labour (Roy 2015).

When scholars such as Shultz (1990) and Shalev (1989) wrote in support of surrogacy, even advocating commercialization of surrogacy (Arneson 1992), it had not globalized to the extent it has today. The impact of commodification of the human body, especially of women's reproductive body parts, through liberalization and globalization is a lot more evident in recent days. The manifestations of objectification in the surrogacy arrangements include instrumentality, denial of subjectivity, inertness and exchangeability (Berkhout 2008), and there is evidence of such forms of objectification of surrogate mothers in India (Saravanan 2013; SAMA 2012; Pande 2010; Vora 2013; Deomampo 2013). Accordingly, academics, activists, lawyers, policymakers, and medical practitioners all over the globe have critiqued the surrogacy practices in India. The recent prohibition of commercial surrogacy practice in India has followed recommendations from all these disciplines within India. The manifestation of commodification such as geneticization, racism, classism, neocolonialism will be examined in the selection criteria, motivation of participants, rhetoric, relationship expressed towards the child, the relinquishment experienced by the intended parents and surrogate mothers in the surrogacy process in Chaps. 5 and 6. Widdows (2009) suggests that we should not be supporting potentially commodifying practices especially when there are ethically pressing reasons to do so (such as in organ transplantation) because the commodification of the body parts gets normalized into a commodification of the persons. A separation of parts and persons should be avoided wherever unnecessary; for instance, in choosing parts and traits in the IVF procedures, concerns regarding commodification at the social level should take precedence over the preferences of intended parents (Widdows 2009).

A feminist perspective needs to include commitments to human rationality along with individual autonomy and to understand the social context of personal choices. Thus, questions of individual reproductive freedom need to be raised in conjunction with human progress which is required for a just society (Ryan 1990). Feminism stands for individual reproductive rights that come along with responsibilities towards a just and humane society. Hence, procreative liberty achieved by violating women's bodily integrity and overlooking mutual human fellowship cannot be considered as an individual "right". According to Raymond (1993), rights need to address power imbalance, justice, self-determination, and international relations which should in turn be grounded in dignity of the individual and integrity of relations between individuals and groups in society.

Stratified Reproduction

Shellee Colen defines stratified reproduction as the physical and social reproductive tasks that "are accomplished differentially according to inequalities that are based on hierarchies of class, race, ethnicity, gender, place in a global economy, and migration status and that are structured by social, economic, and political forces" (Colen 1995: 380). Stratified reproduction is a concept developed by Shellee Colen (1995) to understand the stratification of reproductive labour differentially experienced according to the inequalities of access to resources and socio-economic and political hierarchies. In Colen's research on West Indian childcare workers employed in New York, women (generally of colour) confronted with lesser job opportunities due to their lower education; with the rising cost of living, they chose to migrate for childcare work. They migrated to another country to care for other people's household and children while leaving their children behind. Until they obtained legal status, they were required to work as live-in workers which meant exploitation and contractual conditions without medical insurance and leave provision. Children of these foster mothers felt a sense of confusion, loss, and resentment due to complete absence of their mother and some were left in hostels. With a surplus of women willing to join this work, their reproductive labour was devalued and trivialized as unskilled and newspaper advertisements would contain headings such as "Rent-a-Wife" and "at-your-command" housework. The employers need her services only for childcare, they don't want her to become the mother, although she feels like one especially because she has left their own children behind in another country. Her research reveals that stratification of reproduction along with increasing commodification has reinforced and intensified inequalities of race, class, gender, and migration (Colen 1995). This is very similar to the commercial surrogacy market situation in India; women from poverty ridden households with lesser job opportunities on account of their lower skills, enter into unjust contracts that expect them to live in surrogate homes under controlled monitored conditions consenting to exploitative clauses so that their children can have a better life. There has been a surplus of women willing to participate in surrogacy due to poverty. According to official figures 276 million (22%) of the total population in India presently live below the poverty line (NSSO 2013). The McKinsey Global Institute (MGI) developed a revised analytic index, the "Empowerment line" which estimated that 680 million (56%) of the population lack the means to meet their essential needs (Gupta et al. 2014).

Methodological Framework

Colen's "stratified reproduction" as a conceptual framework, examines the division of reproductive labour that may be differentially experienced according to inequalities of access to resources structured by hierarchies, aspects of which promote or interfere with socio-economic and political status. The concept examines reproductive labour of bearing, rearing, and socializing children that may be differentially experienced, valued, and rewarded according to inequalities of access to material and social resources structured by hierarchies of class, race, ethnicity, gender, place, and migration status differences, aspects of which promote or interfere with socio-economic and political status (Colen 1995). Stratified reproduction refers to the economics, policies and practices that promote and enable reproduction and childrearing for some, while discouraging the same for others based on inequalities. By allocating resources unequally effecting differential experience, value, and reward such stratification of reproductive practices give precedence to the privileged, while making the same reproduction for the under-privileged difficult and even dangerous.

Many feminist social scientists have referred to stratified reproduction (Ikemoto 2015), while some have partially applied it as an analytical approach to examine social reproductive tasks vary based on class, race, ethnic, and global hierarchies (Mohapatra 2012), or to understand the stratification of reproduction (Deomampo 2013). There are no empirical studies that have examined assisted reproductive technology stratified reproduction as a comprehensive methodological framework. Combined with ethnographic research as an analytical framework, the research links a phenomenon such as surrogacy to the everyday structures of racialization and histories inequalities. Stratified reproduction helps to examine the intersectionality in the systems of gender, race, class, and other forms of inequalities. It however, is limited to understanding the differences in experience and power relationships. Hence, this research applies reproductive justice towards identifying a strategic pathway. In this book, I draw on Colen's (1995) stratified reproduction as a framework to examine reproductive justice in the Indian surrogacy bazaar. The objectives of this research were to understand how stratified reproductive tasks of bearing and rearing children is differentially embodied as experience, value, and reward according to inequalities to examine whether this embodiment structured by socio-economic and political status reinforce inequalities and prejudice from a reproductive justice perspective and to identify strategic pathways using intersectionality towards achieving reproductive justice in surrogacy practices. Intersectionality as a methodological framework examines interaction that reveal inequalities but as an analytic framework it is limited to describing the interactions and falls short of examining strategies within these interactions. As one of the historic forms of inequalities and exploitation has been colonization; this book has one chapter on postcolonial theories.

Reproductive Justice, Transnational Feminism, and Intersectionality

Reproductive justice framework is a relatively recent and yet compelling approach in understanding the complex macro and micro level intersectionality of assisted reproductive technologies. A reproductive justice framework includes an analysis of the socio economic, political, and historic contexts within which surrogacy operates. Reproductive justice is a social justice-based framework that links women's reproductive and sexual health and their social, cultural, economic, and political conditions, taking into account the intersecting socio-economic factors, race, religion, geography, nationality, sexuality, and overall health aspects that effect women's lives. Social justice related to peace, poverty, human rights, prejudice and discrimination, educational equity and health and health care inequalities are fundamental to the achievement of reproductive justice (Chrisler 2013–14). The reproductive justice movement has been influenced by global feminism and transnational feminist movements (Morgan 2003; Mohanty 2003). Global feminism works towards women around the globe aiming to solve systemic problems resulting from patriarchy affecting women's health and well-being (Morgan 2003). While reproductive rights can also resist authority to being told what one can or cannot do with one's body, reproductive justice is positive rights that is based on the role of the authorities to support one's pursuit of a good quality of life (Bristow). Transnational feminism identifies the difficulties in dividing women especially in a globalized world influenced by media, multinational corporations, and migration (Mohanty 2003). Transnational feminism also emphasizes on intersectionality of oppression in the form of social class and identity (age, race, ethnicity, socio-economic class, sexual orientation, gender identity, religion, and ability) that has an effect on women's experience. Transnational movements work towards supporting women who develop solutions within their cultural context rather than a hierarchical approach of women from global North attempting to devise solutions for the global South. Intersectionality is a concept coined by Crenshaw (1991), a prominent scholar of critical race theory as a framework to understand intersecting forms of oppression, domination, or discrimination. It is widely applied in a range of feminist analysis, philosophical, and political science research. This approach focuses on ways in which relationships between social structures and systems of oppression combines to produce marginalizing effects. Intersectionality framework attends to context specific inquiries such as; "analysing multiple ways in which race and gender interact with class in the labor market" (Cho et al. 2013: 785).

Raymond (1993) has made reference to 'justice' in the last chapter of her book 'Women as Wombs' in explaining that women's individual bodily dignity and the integrity between individuals and groups in society should be important considerations in determining international human rights. Similar reference to bodily integrity and structural inequalities has also been made from a global gender justice perspective (Donchin 2010). Using a social justice approach, Callahan and Roberts (1996) oppose paid pregnancy contracts as it contributes to subordination of women, poor, and people of colour. Bailey (2011) critiques liberal feminism for extending Western frameworks of liberty to Indian contract mothers and criticizes feminist biomedical ethnologists for weak moral absenteeism resulting in under-theorizing structural harms and injustices. The shortcoming of her work as she herself claims is that she has not included the perspectives of radical feminists. In the context of surrogacy practices in India she suggests that in order to theorize "reproductive justice" the starting point should be to understand the deep injustices that emerge from the surrogacy-or-poverty dilemmas that compel women to take on surrogacy. A key principle of a reproductive justice model is to bring to center the vulnerable people; the poor, people of color people with disabilities and people with non-normative gender expression and sexualities (Luna and Luker 2013). However, it is important to note that reproductive exploitation can occur both within and between these vulnerable groups.

Asian Communities for Reproductive Justice (ACRJ) has developed three frameworks on reproductive health, rights, and justice. The Reproductive Justice framework includes recognizing the histories of reproductive oppression in all communities. This model is based on organizing women/girls to change structural power inequalities. It examines the control and exploitation of women's bodies, sexuality and reproduction as it has been used as an effective strategy for controlling women and communities, particularly those of colour which is manifested through multiple oppressions of race, class, gender, sexuality, ability, age, and immigration status. "Controlling a woman's body controls her life, her options and her potential" (ACRJ 2005: 2). This model is based on the human rights framework published by ACRJ along with the Sister Song Collective with an aim to bring "reproductive justice" into the mainstream of "reproductive rights" and "social justice movements". A step towards this would be to examine where reproductive technologies cross the "humanitarian" threshold of the feminist ideologies of equality, liberty, and justice. These conceptual frameworks will be revisited again in Chaps. 4 and 5, working towards identifying strategic pathways of humanitarian assisted conception based on the empirical findings.

References

ACRJ. 2005. A new vision for advancing our movement for reproductive health, reproductive rights and reproductive justice. Oakland, CA: Asian Communities for Reproductive Justice, Website.

Agamben, G. 1998. Homo sacer: Sovereign power and bare life. Translated by D. Heller-Roazen. Stanford: Stanford University Press.

Alcoff, L. 2006. *Visible identities: Race, gender, and the self*. New York: Oxford University Press.

Anderson, Elizabeth S. 1990. Is women's labor a commodity? *Philosophy & Public Affairs* 19 (1): 71–92.

Anderson, Elizabeth S. 2000. Why commercial surrogate motherhood unethically commodifies women and children: Reply to McLachlan and Swales. *Health Care Analysis* 8: 19–26.

Andrews, Lori. B. 1986. The stork market: The law of the new reproduction …the baby selling laws. *Columbia Journal of Law and Social Problems* 20: 1–53.

Andrews, Lori. 1989. *Between strangers: Surrogate mothers, expectant fathers, and brave new babies*. New York: Harper & Row.

Appadurai, A. 1986. *The social life of things: Commodities in cultural perspective*. Cambridge: Cambridge University Press.

Arlamovsky, Maria. 2016. 'Future baby', NGF-Nikolaus Geyrhalter Filmproduktion.

Arneson, Richard J. 1992. Commodification and commercial surrogacy. *Philosophy and Public Affairs* 21 (2): 132–16.

Atwood, Margaret. 1985. *The handmaid's tale*. Toronto: McClelland & Stewart.

Bailey, Alison. 2011. Reconceiving surrogacy: Toward a reproductive justice account of Indian surrogacy. *Hypatia* 26: 715–741.

Baker, Brenda M. 1996. A case for permitting altruistic surrogacy. *Hypatia* 11: 34–48.

Banerjee, Amrita. 2011. Re-conceiving "borders": A feminist pragmatic phenomenology for postcolonial feminist ethics and politics. A dissertation Presented to the Department of Philosophy and the Graduate School of the University of Oregon.

Becker, G. 2000. *The elusive embryo: How women and men approach new reproductive technologies*. Berkeley: University of California Press.

Bentham, Jeremy. 1982. *Introduction to the principles of morals and legislation*. London: Methuen.

Berend, Zsuzsa. 2010. Surrogate losses: Understandings of pregnancy loss and assisted reproduction among contract mothers. *Medical Anthropology Quarterly* 24: 240–262.

Berkhout, Suze G. 2008. Buns in an oven: Objectification, surrogacy and women's autonomy. *Social Theory and Practice* 34: 95–117.

Bharadwaj, A. 2016. *Conceptions: Infertility and procreative technologies in India*. New York, Oxford: Berghahn.

Blyth, E. 1994. I wanted to be interesting. I wanted to be able to say 'I've done something interesting with my life: Interviews with SMs in Britain. *Journal of Reproductive and Infant Psychology* 12 (3): 189–198.

Bordo, S. 1993. *Unbearable weight: Feminism, western culture and the body*. Berkeley: University of California Press.

Brähler, E., and Y. Stöbel-Richter. 2002. Soziologische, epidemiologische und politische Aspekte der Reproduktionsmedizin. In *Vom Stammbaum zur Stammzelle. Reproduktionsmedizin, Pränataldiagnostik und menschlicher Rohstoff (S. 97-110)*, ed. E. Brähler, Y. Stöbel-Richter und U. Hauffe. Gießen: Psychosozial-Verlag.

Callahan, Joan C., and Dorothy E. Roberts. 1996. A feminist social justice approach to reproduction-assisting technologies: A case study on the limits of liberal theory. *Faculty Scholarship,* Paper 1155.

Carmeli, Yoram S., and Daphna Birenbaum-Carmeli. 1994. The predicament of masculinity: Towards understanding the male's experience of infertility treatment. *Sex Roles* 30 (9–10): 663–677.

Cho, Sumi, Kimberlé Williams Crenshaw, and Leslie McCall. 2013. Toward a field of intersectionality studies: Theory, applications, and praxis. *Signs* 38 (4): 785–810.

Chrisler, Joan C. 2013. Introduction: A global approach to reproductive justice psychosocial and legal aspects and implications. *William & Mary Journal of Women and the Law* 20: 1–25.

Colen, S. 1995. Like a mother to them: Stratified reproduction and West Indian childcare workers and employers in New York. In *Conceiving the new world order: The global politics of reproduction*, ed. F. Ginsburg and R. Rapp, 78–102. Berkeley: University of California Press.

Corea, Gena. 1985. *The mother machine: Reproductive technologies from artificial insemination to artificial wombs*. New York: Harper and Row.

Cornell, Drucilla. 1998. *At the heart offreedom: Feminism, sex, and equality*. Princeton: Princeton University Press.

Crenshaw, Kimberle. 1991. Mapping the margins: Intersectionality, identity politics, and violence against women of color. *Stanford Law Review* 43 (6): 1241–1299.

DasGupta, Sayantani, and Shamita Das DasGupta. 2014. *Globalization and transnational surrogacy in India: Outsourcing life*. Lanham: Lexington Books.

Dawe, Gavin S., W. Tan Xiao, and Xiao Zhi-Cheng. 2007. Cell migration from baby to mother. *Cell Adhesion and Migration* 1: 19–27.

Deomampo, Daisy. 2013. Gendered geographies of reproductive tourism. *Gender & Society* 27: 514–537.

Dickenson, Donna. 2007. *Property in the body: Feminist perspectives*. Cambridge: Cambridge University Press.

Donchin, A. 1996. Feminist critiques of new fertility technologies: Implications for social policy. *The Journal of Medicine and Philosophy* 21 (5): 475–498.

Donchin, A. 2010. Reproductive tourism and the quest for global gender justice. *Bioethics* 24 (7): 323–332.

Douglas, M. 1970. *Natural symbols: Explorations in cosmology*. London: Routledge.

Dube, L. 1998. On the construction of gender: Hindu girls in patrilineal India. In *Socialization, education and Women: Explorations in gender identity*, ed. K. Chanana, 168–192. New Delhi: Orient Longman.

Dworkin, Andrea. 1983. *Right-wing women*. New York: Perigee Books.

Ehrich, K., and C. Williams. 2010. A 'healthy baby': The double imperative of preimplantation genetic diagnosis. *Health* 14: 41–56.

Finley, Nancy J. 1989. Theories of family labor as applied to gender differences in caregiving for elderly parents. *Journal of Marriage and Family* 51 (1): 79–86.

Foucault, M. 1997. *Ethics, subjectivity and truth: The essential works of Foucault*, ed. Paul Rabinow. Translated by Robert Hurley and others.

Franklin, S., and C. Roberts. 2006. *Born and made: An ethnography of preimplantation genetic diagnosis*. New Jersey, Oxfordshire: Princeton University Press.

Fulfer, Katy. 2017. Cross-border reproductive travel, neocolonialism, and Canadian policy. *International Journal of Feminist Approaches to Bioethics* 10 (1): 225–247.

Ginsburg, F., and R. Rapp. 1995. *Conceiving the new world order: The global politics of reproduction*. Berkeley: University of California Press.

Greil, A.L. 1991. *Not yet pregnant: Infertile couples in contemporary America*. New Brunswick: Rutgers University Press.

Greil, A.L., K. Slauson-Blevins, and J. McQuillan. 2010. The experience of infertility: A review of recent literature. *Sociology of Health & Illness* 32: 140–162.

Greil, A.L., J. McQuillan, M. Lowry, and K.M. Shreffler. 2011a. Infertility treatment and fertility-specific distress: A longitudinal analysis of a population-based sample of U.S. women. *Social Science and Medicine* 73 (1): 87–94.

Greil, Arthur, Julia McQuillan, and Kathleen Slauson-Blevins. 2011b. The social construction of infertility. *Sociology Compass* 5 (8): 736–746.

Gupta, J.A. 2000. *New reproductive technologies, women's health and autonomy: Freedom or dependency*. New Delhi: Thousand Oaks.

Gupta, R., S. Sankhe, R. Dobbs, J. Woetzel, A. Madgavkar, and A. Hasyagar. 2014. *From poverty to empowerment: India's imperative for jobs, growth and effective basic services*. Mumbai: McKinsey Global Institute.

Gutmann, M.C. 2007. *Fixing men: Sex, birth control, and AIDS in Mexico*. Berkeley: University of California Press.

Harrison, Laura. 2016. *Brown bodies, white babies: The politics of cross-racial surrogacy*. New York: New York University Press.

Hassan, N. 2010. Milk markets: Technology, the lactating body, and new forms of consumption. *WSQ: Women's Studies Quarterly* 38 (3): 209–228.

Held, Virginia. 1987. Non-contractual society: A feminist view. In *Science, morality, and feminist theory*, ed. M.P. Hanen and K. Nielsen. Calgary: University of Calgary Press.

Hill, John Lawrence. 1990. The case for enforcement of the surrogate contract. *Politics and the Life Sciences* 8 (2): 147–160.

Hochschild, A. R. 2004. The commodity frontier. In *Self, social structure and beliefs: Essays in sociology*, ed. J. Alexander, G. Marx, and C. Williams. Berkeley: University of California.

Hochschild, Arlie. 2011. Emotional life on the market frontier. *The Annual Review of Sociology* 37: 21–33.

ICPD Programme of Action. 1994. *Cairo, Egypt: International Conference on Population and Development*. United Nations Population Funds.

Ikemoto, Lisa C. 2015. Egg freezing, stratified reproduction and the logic of not. *Journal of Law and the Biosciences* 1 (4): 112–117.

ILO. 2012. *Statistical update on employment in the informal economy ILO*. Department of Statistics. http://laborsta.ilo.org/applv8/data/INFORMAL_ECONOMY/2012-06-Statistical%20update%20-%20v2.pdf. Accessed 15 Aug 2017.

Inhorn, Marcia C. 2002. Sexuality, masculinity, and infertility in Egypt: Potent troubles in the marital and medical encounters. *The Journal of Men's Studies* 10: 343–359.

Inhorn, Marcia C. 2009. Right to assisted reproductive technology: Overcoming infertility in low-resource countries. *International Journal of Gynecology and Obstetrics* 106: 172–174.

Inhorn, M., and P. Patrizio. 2012. The global landscape of cross-border reproductive care: Twenty key findings for the new millennium. *Current Opinion in Obstetrics and Gynecology* 24: 158–163.

Inhorn, Marcia C., and Frank van Balen. 2002. *Infertility around the globe: New thinking on childlessness, gender, and reproductive technologies*. Berkeley: University of California Press.

Jaggar, A.M., and P.R. Struhl. 1978. *Feminist frameworks: Alternative theoretical accounts of the relations between women and men*. New York: McGraw-Hill.

Johnson, Katherine M., and Jasmine Fledderjohann. 2012. Revisiting "her" infertility: Medicalized embodiment, self-identification and distress. *Social Science and Medicine* 75 (5): 883–891.

Katz, Avi. 1986. Contract motherhood and the baby-selling laws. *Columbia Journal of Law and Social Problems* 20: 1–53.

Khetarpal, Abha, and Satendra Singh. 2012. Infertility: Why can't we classify this inability as disability? *Australasian Medical Journal* 5 (6): 334–339.

Kimbrell, A. 1993. *The human body shop the engineering and marketing of life*.

Kishwar, M. 1993. Abortion of female fetuses: Is legislation the answer? *Reproductive Health Matters* 1 (2): 113–115.

Klein, Renate. 2017. *Surrogacy: A human rights violation*. Melbourne: Spinifex Press.

Krones, T., and G. Richter. 2004. Preimplantation genetic diagnosis (PGD): European perspectives and the German situation. *Journal of Medicine and Philosophy* 29: 623–640.

Kuchroo, Sakshi. 2016. Banning commercial surrogacy is the only way forward: Professor Mohan Rao of Jawaharlal Nehru University talks about the pros and cons of the surrogacy bill. *Governance Now* 7 (16): 38–39.

Lamont, M'le, S. Pendergrass, and M. Pachucki. 2015. Symbolic boundaries. *International Encyclopedia of the Social & Behavioral Sciences* 23: 850–55.

Letherby, Gayle. 2002. Challenging dominant discourses: Identity and change and the experience of 'infertility' and 'involuntary childlessness'. *Journal of Gender Studies* 11 (3).

Littleton, Christine A. 1987. Reconstructing sexual equality. *California Law Review* 1279. http://scholarship.law.berkeley.edu/californialawreview/vol75/iss4/2. Accessed 15 Aug 2017.

Luna, Zakiya, and Kristin Luker. 2013. Reproductive justice. *Annual Review of Law and Social Science* 9: 327–352.

MacKinnon, Catharine A. 2011. Trafficking, prostitution, and inequality. *Harvard Civil Rights-Civil Liberties Law Review* 46: 271–309.

Majumdar, Anindita. 2014. The rhetoric of choice: The feminist debates on reproductive choice in the commercial surrogacy arrangement in India. *Gender, Technology and Development* 18 (2): 275–301.

Mauss, M. 1985. A category of the human mind: The notion of the person; the notion of the self. *The category of the person* 1–25.

Merleau-Ponty, M. 1995 (1945). *Phenomenology of perception*. Translated by Colin Smith. New York: Routledge.

Miall, Charlene E. 1986. The stigma of involuntary childlessness. *Social Problems* 33 (4): 268–282.

Mies, M. 1988. From the individual to the dividual: In the supermarket of "reproductive alternatives". *Reproductive and Genetic Engineering* 1 (3): 225–237.

Mill, J.H. 1989. Chapter on socialism. In *On liberty and other writings*, ed. Stefan Collini. Cambridge: Cambridge University Press.

Mishra, K., and A. Dubey. 2014. Indian women's perspectives on reproduction and childlessness: Narrative analysis. *International Journal of Humanities and Social Science* 4.

Mistry, R., O. Galal, and M. Lu. 2009. Women's autonomy and pregnancy care in rural India: A contextual analysis. *Social Science and Medicine* 69 (6): 926–933.

Mohanty, Chandra Talpade. 2003. Under western eyes" revisited: Feminist solidarity through anticapitalist struggles. *Signs* 28 (2): 499–535.

Mohapatra, Seema. 2012. Achieving reproductive justice in the international surrogacy market. *Annals of Health Law* 21 (1): 191–200.

Morgan, R. (1996). *Sisterhood is global*. New York: Feminist Press.

Morgan, Robin. 2003. *Sisterhood is forever: The women's anthology for a new millennium*. New York: Washington Square Press.

Nahar, P., and A. Richters. 2011. Suffering of childless women in Bangladesh: The intersection of social identities of gender and class. *Anthropology and Medicine* 18 (3): 327–338.

NDTV. 2015. We the people—The surrogacy debate; of motherhood or money? *YouTube*. https://www.youtube.com/watch?v=7OlBHVKZzjg. Accessed 14th Aug 2017.

NSSO. 2012. *Informal sector and conditions of employment in India NSS 66th ROUND* (July 2009–June 2010). National Sample Survey Office National Statistical Organisation Ministry of Statistics & Programme Implementation Government of India.

NSSO. 2013. *Key indicators of employment and unemployment in India*. New Delhi: The National Sample Survey Office (NSSO), Ministry of Statistics and Programme Implementation.

Padmanabhan, Anil. 2016. Formalizing India's informal economy. *Live Mint*. http://www.livemint.com/Opinion/4kxau3CejBat9uAIEakIKK/Formalizing-Indias-informal-economy.html . Accessed 15 Aug 2017.

Pande, A. 2010. Commercial surrogacy in India: Manufacturing a perfect mother-worker. *Signs* 35: 969–992.

Pande, A. 2014. *Wombs in labor: Transnational commercial surrogacy in India*. New York: Columbia University Press.

Pateman, Carole. 1988. *The sexual contract*. Stanford: Stanford University Press.

Petchesky, Rosalind Pollack. 1995. From population control to reproductive rights: Feminist fault lines. *Reproductive Health Matters* 3 (6): 152–161.

Pfeffer, N. 2011. Eggs-ploiting women: A critical feminist analysis of the different principles in transplant and fertility tourism. *Reproductive Biomedicine Online* 23 (5): 634–641.

Philip, C.M. 2017. *Aadhar helps railways trace parents of trafficked kids*. Bangaluru: The Times of India, 8th November.

Pujari, S., and S. Unisa. 2014. Failing fatherhood: A study of childless men in rural Andhra Pradesh. *Sociological Bulletin* 63: 21–40.

Radin, M.J. 1995. What, if anything, is wrong with baby selling? *Pacific Law Journal*.

Rapp, Rayna. 1999. *Testing women, testing the fetus: The social impact of amniocentesis in America*, 368. New York: Routledge.

Raymond, Janice G. 1990. Reproductive gifts and gift giving: The altruistic woman. *Hastings Center Report* 20 (6): 7–11.

Raymond, J.G. 1993. *Women as wombs: Reproductive technologies and the battle over women's freedom*. San Francisco: Harper.

Reddy, Sunita, and Tulsi Patel. 2015. There are many eggs in my body: Medical markets and commodified bodies in India. *Global Bioethics* 26 (3–4): 218–231.

Riessman, C. 2000. Stigma and everyday resistance practices, childless women in South India. *Gender Society* 14: 111–135.

Roberts, E.F. 1998. "Native" narratives of connectedness: Surrogate motherhood and technology. In *Cyborg babies: From techno-sex to techno-tots*, ed. R. Davis-Floyd and J. Dumit, 193–211. New York: Routledge.

Roberts, Dorothy E. 1996. Race and the new reproduction. *Faculty Scholarship*. Paper 1154.

Robertson, John A. 1983. Procreative liberty and the control of conception, pregnancy, and childbirth. *Virginia Law Review* 69 (3): 405–464.

Robertson, John A. 1986. Embryos, families, and procreative liberty: The legal structure of the new reproduction. *Southern California Law Review* 59: 939–1041.

Rose, Nikolas. 2001. The politics of life itself. *Theory, Culture and Society* 18 (6): 1–30.

Rotabi, K.S., N.F. Bromfield, and P. Fronek. 2015. International private law to regulate commercial global surrogacy practices: Just what are social work's practical policy recommendations? *International Social Work*. https://doi.org/10.1177/0020872814564706.

Rothenberg, K.H. 1996. Feminism, law, and bioethics. *Kennedy Institute of Ethics Journal* 6 (1): 69–84.

Rothman, Barbara Katz. 1992. Reproductive technologies and surrogacy: A feminist perspective. *Creighton Law Review* 25: 1599–1616.

Roy, S. 2015. *Trafficked tribal girls forced to conceive, deliver babies for sale*. New Delhi: Hindustan Times: 25th Feb.

Rudrappa, S., and C. Collins. 2015. Altruistic agencies and compassionate consumers: Moral framing of transnational surrogacy. *Gender & Society* 29: 937–959.

Ryan, Maura A. 1990. The argument for unlimited procreative liberty: A feminist critique. *Hastings Center Report* 20: 6–12.

SAMA. 2012. *Birthing a market: A study on commercial surrogacy*. New Delhi: SAMA-resource group for women and health.

Sandelowski, M. 1991. Compelled to try: The never-enough quality of conceptive technology. *Medical Anthropology Quarterly* 5: 29–47.

Sandelowski, Margarete, Diane Holditch-Davis, and Betty G. Harris. 1990. Living the life: Explanations of infertility. *Sociology of Health & Illness* 12 (2): 195–215.

Saravanan, S. 2010. Transnational surrogacy and objectification of gestational mothers. *Economic and Political Weekly* 45: 26–29.

Saravanan, S., and R.R. Ranadive. 2010. *Mother anonymous (Documentary)*. India: Frame of Mind Communications Pvt. Ltd. Mumbai and Asia and Europe in a Global Context, University of Heidelberg, Germany.

Saravanan, S. 2013. An ethnomethodological approach to examine exploitation in the context of capacity, trust and experience of commercial surrogacy in India. *Philosophy, Ethics and Humanities in Medicine* 8: 10.

Saravanan, S. 2015. Global justice, capabilities approach and commercial surrogacy in India. *Medicine, Health Care and Philosophy* 18: 295–307.

Saravanan, S. 2016. 'Humanitarian' thresholds of the fundamental feminist ideologies: Evidence from surrogacy arrangements in India. *Analize: Journal of Gender and Feminist Studies* 6: 66–88.

Saravanan, S. 2017. When doctors rock the cradle. *Governance Now* 8: 56–58.

Sayeed, U. 1999. Childlessness in Andhra Pradesh, India: Treatment seeking and consequences. *Reproductive Health Matters* 7 (13): 54–64.

Scheman, Naomi. 1983. Individualism and the objects of psychology. *Discovering Reality* 161: 225–244.

Scheper-Hughes, Nancy. 2000. The global traffic in human organs. *Current Anthropology* 41 (2): 191–224.

Scheper-Hughes, N., and M. Lock. 1987. The mindful body: A prolegomenon to future work in medical anthropology. *Medical Anthropology Quarterly* 1: 6–41.

Schick, M., S. Rösner, B. Toth, T. Strowitzki, and T. Wischmann. 2016. Exploring involuntary childlessness in men—A qualitative study assessing quality of life, role aspects and control beliefs in men's perception of the fertility treatment process. *Human Fertility* 19 (1): 32–42.

Schroeder, P. 1988. Infertility and the world outside. *Fertility and Sterility* 49: 765–767.

Senarath, U., and N.S. Gunawardena. 2009. Women's autonomy in decision making for health care in South Asia. *Asia-Pacific Journal of Public Health* 21 (2): 137–143.

Shalev, Carmel. 1989. *Birth power: The case for surrogacy*. New Haven: Yale University Press.

Shalev, C. 1998. Halakha and patriarchal motherhood - An anatomy of the new Israeli Surrogacy Law. 32. *Israeli Law Review*: 51.

Shanley, Mary Lyndon. 1993. "Surrogate mothering" and women's freedom: A critique of contracts for human reproduction. *Signs* 18 (3): 618–639.

Sharp, L.A. 2000. The commodification of the body and its parts. *Annual Review of Anthropology* 1: 287–328.

Shultz, Marjorie M. 1990. Reproductive technology and intention-based parenthood: An opportunity for gender neutrality. *Wisconsin Law Review* 297–398.

Sloan, C., B. Gough, and M. Conner. 2010. Healthy masculinities? How ostensibly healthy men talk about lifestyle, health and gender. *Psychology and Health* 25: 783–803.

Stöbel-Richter, Y., M.E. Beutel, C. Finck, and E. Brähler. 2005. The 'wish to have a child', childlessness and infertility in Germany. *Human Reproduction* 20: 2850–2857.

Subedi, Madhusudan. 2015. Contractual transaction: How renting a uterus makes the human body a commodity in Nepal. *Dhaulagiri Journal of Sociology and Anthropology* 9: 1–25.

Swazey, J.P. 1992. *Spare parts: Organ replacement in American Society*. Oxford: Oxford University Press.

Teman, Elly. 2009. Embodying contract motherhood: Pregnancy as a dyadic body-project. *Body & Society* 15: 47–69.

Teman, Elly. 2010. *Birthing a mother: The surrogate body and the pregnant self*. Berkeley: University of California Press.

Thompson, Hazel. 2013. Horrors of India's brothels documented. *British Broad Casting*. http://www.bbc.com/news/world-asia-india-24530198. Accessed 15 Aug 2017.

Tieu, Mathew M. 2009. Altruistic surrogacy: The necessary objectification of contract mothers. *Journal of Medical Ethics* 35: 171–175.

Tong, Rosemarie. 2009. *Feminist thought: A more comprehensive introduction*. Colarado: Westview Press.

Truong, T.D. 2001. Human trafficking and organised crime. *ISS Working Paper Series/General Series* 339:1–30.

UNFPA. 2004. *Investing in people: National progress in implementing the ICPD programme of action 1994–2004*. United Nation Population Fund New York.

Unisa, S. 1999. Childlessness in Andhra Pradesh, India: Treatment-seeking and consequences. *Reproductive Health Matters* 7: 54–64.

Van Balen, F., and H.M.W. Bos. 2009. The social and cultural consequences of being childless in poor-resource areas. *Facts, Views & Vision in ObGyn* 1 (2): 106–121.

Van Niekerk, Anton and Liezl Van Zyl. 1995. The Ethics of surrogacy: Women's reproductive labour. *Journal of Medical Ethics* 21: 345–349.

Vora, Kalindi. 2009. Indian transnational surrogacy and the commodification of vital energy. *Subjectivities*. 28 (1): 266–278.

Vora, Kalindi. 2013. Potential, risk and return in transnational Indian gestational surrogacy. *Current Anthropology* 54: S97–S106.

Vora, Kalindi. 2010. Medicine, markets and the pregnant body: Indian commercial surrogacy and reproductive labor in a transnational frame. *Scholar & FeministOnline* 9(1/2).

Wendell, Susan. 1987. A (qualified) defense of liberal feminism. *Hypatia* 2 (2): 65–93.

Whiteford, L.M., and L. Gonzalez. 1995. Stigma: The hidden burden of infertility. *Social Science and Medicine* 40: 27–36.

Whittaker, Andrea M. 2011. Reproduction opportunists in the new global sex trade: PGD and non-medical sex selection. *Reproductive BioMedicine Online* 23: 609–617.

Wichterich, C. 2000. *The globalized woman: Reports from a future of inequality*. North Melbourne: Spinifex Press.

Widdows, Heather. 2009. Border disputes across bodies: Exploitation in trafficking for prostitution and egg sale for stem cell research. *International Journal of Feminist Approaches to Bioethics* 2 (1): 5–24.

Widge, A. 2002. Sociocultural attitudes towards infertility and assisted reproduction in India. *Current Practices and Controversies in Assisted Reproduction* 60–74.

Williams, Wendy Webster. 1982. The equality crisis: Some reflections on culture, courts, and feminism. *Women's Rights Law Report* 7: 175.

Wirtberg, I., A. Möller, L. Hogström, S.E. Tronstad, and A. Lalos. 2007. Life 20 years after unsuccessful infertility treatment. *Human Reproduction* 22: 598–604.

Wischmann, T., and P. Thorn. 2013. (Male) infertility: What does it mean to men? New evidence from quantitative and qualitative studies. *Reproductive Biomedicine Online* 27: 236–243.

Young, Iris Marion. 2005. On female body experience "throwing like a girl" and other essays. *Feminism & Psychology*. https://doi.org/10.1093/0195161920.003.0001.

Zilberberg, J. 2007. Sex selection and restricting abortion and sex determination. *Bioethics* 21: 517–519.

Part II
Indian Surrogacy Biomarkets

Chapter 4
Situating India in the Globalscape of Inequalities

Abstract India became a popular global destination for assisted reproduction, supported by the Indian government. In the backdrop of this increasing popularity was ineffective governance, unregistered clinics, and a growing market that was embedded in socio-economic inequalities in India. This chapter reviews the inequalities, ineffective governance, unclear surrogacy regulations, and unethical practices which have made countries such as India an ideal environment for global injustice in the process of commercial surrogacy. Women with limited entitlements in terms of education and employment possibilities entered into unjust surrogacy contracts that included triple alienation—from their own bodies, an alienation from the rest of the world, and from the babies born to them. The chapter begins with this backdrop of socio-economic and health inequalities in India as most surrogate mothers are primary or high-school drop-outs before proceeding with the inequalities within the surrogacy biomarkets; unfair distribution of benefits, alienation, violation of good medical practices, and the commodification of children and women's bodies. I also introduced and contested the liberal argument that normalizes surrogacy as any other form of work or labour.

Keywords Socio-economic inequalities · Maternal health · Surrogacy as work
Surrogate homes · Segmented reproduction

India is a popular global health destination, providing medical care, equipment, and facilities at comparatively low costs. A subset of medical tourism is reproductive health care, including treatments such as assisted reproductive technologies (ARTs) and surrogacy. The official registration of in vitro fertilization (IVF) clinics in India is inadequate: many clinics remain unregistered. According to a recent Indian Council of Medical Research (ICMR) report, while there are 385 IVF registered clinics in India, the National Commission for Women, estimates that 3000 clinics offer surrogacy (ICMR 2015; Kannan 2009). Operating under a laissez-faire approach to ARTs, India primarily promotes such treatment within a financial framework and the priorities of the state are not as focused on the public sector as much on the promotion of the interests of industry which allows for medical

S. Saravanan, *A Transnational Feminist View of Surrogacy Biomarkets in India*,
https://doi.org/10.1007/978-981-10-6869-0_4

tourism to become a site for corporate profit. The Indian government promotes ART markets, sometimes under the term medical tourism, and offers "incentives like low interest rates for loans provided to establish hospitals, and subsidized rates for buying drugs, importing equipment, and buying land for clinics" (Sarojini et al. 2011: 3). This led to an increase in foreigners, estimated at 150,000 in 2004, who visit India strictly for access to ARTs which they may be denied in their home country (Sarojini et al. 2011).

In general, anyone can walk into Indian hospitals seeking treatment as there are no identity cards required, while this may change with the introduction of Aadhar cards. A citizen can also cross state borders to access medical treatment. Hence, Non-Resident Indians (NRIs) who enter the country can easily access any hospital for their treatment and often they do so for surrogacy too. Recently, India introduced medical visas for those who enter the country for medical treatment. The actual number of people entering India for medical treatment is unclear. In my study, I observed that most intended parents who enter India for surrogacy come on tourist visas as advised by the clinics, hence none of them are recorded as medical tourists. Data becomes even more complicated because it is not the surrogate mothers but the intended mothers who register as being pregnant in the clinics and people easily produce through documents with the help of the clinics that it was they who delivered a baby in India. The percentage of foreigners involved in surrogacy in India is reported as approximately 60–80% (Frontline 2016).

One of the most popular clinics in Western India claimed the birth of 500 babies in 2013 of whom two-third were for intended parents from abroad and now it is estimated that the birth of more than 1100 babies through surrogacy occurring until 2016 (Bhalla and Thapiyal 2013; NDTV 2015, 2016). According to a recent estimation, the surrogacy business in India accounted for a yearly return of 2.3 billion Dollar (Perappadan 2014). The estimate of people from abroad seeking surrogacy in India varied from one-third to half of the total seekers (Bhalla and Thapiyal 2013; Desai 2012). Until the ban on surrogacy in September 2016, couples from abroad with infertility issues were specifically attracted to India due to its liberal and unregulated laws, the low cost and easy availability of women willing to become surrogate mothers, and the surrogate mothers' limited rights in India compared to those in the USA or Canada. This places surrogate mothers in double vulnerabilities; because of their lower capabilities (socio-economic, educational capacity) and for being women. Furthermore, the ART Bill had nipped in the bud any possible legal agency they could have enjoyed within the surrogacy process; they have no control over their bodies (medical intervention, isolation in surrogate homes, and control over their activities within these homes), the babies born, the kind of relinquishment and relationship with the intended parents. Just because they have no other choice, doesn't mean that this is a justified and dignified form of human labour that the governments in countries like India should encourage. Women as surrogate mothers in India were the nightmare of Corea and Atwood come alive supported by the governance. Fortunately, the Government of India realized this triple alienation of women and put a halt on the commodification on women's bodies by banning commercial surrogacy.

Socio-economic Inequalities and Surrogacy in India

India had been experiencing a boom as a destination for surrogacy services, although the country retains a relatively low global rank in terms of human development and gender equality. India ranks approximately 135 on worldwide human development indicators (HDIs), and other countries commonly that have been providing surrogacy services also have relatively lower HDI ranks (e.g. Nepal 145, Ukraine 83, Republic of Georgia 79, and Mexico 71). Poverty in India has reduced over time due to an increasing public spending and poverty eradication programmes. However, according to official figures, 267 million (22%) of the total population in India presently live below the poverty line (NSSO 2013). The McKinsey Global Institute (MGI) developed a revised analytic index, the "Empowerment line" which estimated that 680 million (56%) people lack the means to meet their essential needs (Gupta et al. 2014). India has also recently started a welfare programme of a cash pay-out for the poor. Gupta and others reported that only an estimated half of the total public money spent on basic services actually reached the beneficiaries with much of it lost to inefficiency or corruption (Gupta et al. 2014). Without reforms in political will, ineffectiveness of governance would constrain future impacts of public spending resulting in a very slow decline of poverty in India.

Poorer women are involved in surrogacy in India and it is one of the main reasons that it is cheaper compared to the USA, UK, Canada, and Australia. In the USA, the costs of commissioning a surrogacy through an agency accounts to a minimum of 200,000 USD, whereas in India it would cost approximately 50,000 USD including travelling and living expenditure (SSA). Other sources say that in Canada an altruistic surrogacy costs almost as much (80,000 CND) as the total cost of surrogacy in India (Fertility Consultants Canada 2014). There is also a difference in the portion of payment received by surrogate mothers in India compared with their counterparts abroad. For example, in the United States a surrogate mother may receive up to 35% of the total surrogacy cost, whereas in India a surrogate mother is paid only 15–25% of the total costs (Saravanan 2013). Even the highest paying clinic that imposes mandatory rules on women to stay in "surrogate homes" pays only up to 30% of the total cost (Saravanan 2015).

In the documentary film *Made in India*, the intended parents were paying about 7,000 dollars for the surrogate mother to carry the baby, but the surrogate mother was paid only 2,000 USD. Dr. Allahabadia in the documentary says he has no idea about the details in the contract between the medical tourism agency and the parents, but was he ever concerned enough to find out? Hence, the intended parents were over-charged and the surrogate mothers were not paid the full amount that was taken on their behalf. The medical practitioner was unaware, disinterested, or acting naïve, but and yet a part of this deal. In my study without any role of an international agent, the situation was similar as there was a huge difference between what was asked from the intended parents on behalf of the surrogate mothers and what was actually paid to her. The surrogate mothers were categorically told not to discuss money with me. The intended parents were overcharged, hence exploited on fees as they are not provided with complete information about the possible extra costs they would incur on caesarean sections; for intensive neonatal care in hospitals; towards making official documents; and for surrogate mothers who would continue to stay in the surrogate homes, providing breastmilk as well as nanny care to the new born (Saravanan 2015). Additional cost is also requested on the pretext of paying surrogate mothers for breastfeeding. It is known that all deliveries are invariably caesarean sections, and not all surrogate mothers happily consent to staying in the surrogate homes after delivery or in providing breastmilk. It is also known that intended parents are charged double for every extra child (refer Chaps. 5 and 6). Additionally, travel costs, hotel charges, bonuses for surrogate mothers add up almost to what the intended parents would spend in Canada, but in India they are ensured to have the baby without any legal hassles.

One intended mother from the USA sums up the motivation, "in New York, the contract is not binding, so beyond going through all the paper work, if the surrogate decided she wanted to keep the baby, the contract is void. She has all the rights"[1] (The VJ Movement 2009). So, regardless of the cost difference, if the surrogate mothers were empowered with more rights in India, this intended mother would not have come here from Canada. Additionally, surrogate mothers in India have to sign off their rights over their own body for the surrogacy period; they have no say on the medical interventions, such as selective abortions, number of embryos to be transferred, they have to stay in dormitory homes, abide by the rules within the home relating to what to watch on TV, what kind of music to listen, they cannot go out for walks, they are not allowed any form of exercise as it may cause exertion and miscarriage. They have to breastfeed and become nannies to the children, and finally relinquish them without a sigh for some bonus money. Thus, surrogate mothers in India often enter into unjust surrogacy contracts owing to their reduced power to negotiate (Saravanan 2015).

Most of the women serving as surrogate mothers in India, do so to provide their family with immediate basic human needs, to "make ends meet" by providing adequate food, while some are living on the edge of poverty wanting to provide for their children's education, to buy a house, and to avoid slipping further into poverty. Other women become surrogate mothers to repay debts, or financial crises that arose due to illness or disability within the family (Saravanan 2015). Women as surrogate mothers in India are from the lower socio-economic quartile, hence, it is important to get a glance of the context of these inequalities in India and place this in the globalscape. Although overall inequalities in India are reducing over a period of time, great differences continue to remain.

Although malnutrition is slowly declining, nearly half of India's children under 3 years of age are malnourished (NFHS 2007). In 2005–2006, more than half of the women in India aged 15–49 years suffered from anaemia (55.3%), an increase of

[1]Incidentally, I was at the surrogate home during the baby shower ceremony when the documentary crew was shooting the film.

3% points over 1998–1999 (NFHS 2007). Maternal mortality rate in India is 167/100,000 live births which is high compared to 12 in more developed regions of the world. The lifetime risk of maternal death is 1 in 370 in India while it is one in 4900 in more developed countries (SRS 2011–2013). According to the doctors, women undergo a health screening before they begin the surrogacy process. However, what is practised in reality is what Dr. Kadam matter-of-factly explained in a documentary title *Made in India*, "I only need her uterus" (Haimowitz and Sinha 2010). Accordingly, the womb is prepared to hold the baby implanted, and other factors are manipulated; women who have a low haemoglobin are pumped with medicine. Another presumption is that all such nutritional or health problems can be prevented if surrogate mothers are provided with the best medical services. One intended mother assured that surrogate mothers in India have nutritionists, they have nurses who watch them every two weeks, they get ultrasounds all the time, hence the fear factor goes down with the extent of health care inputs (Today 2008). Nevertheless, there have been deaths of surrogate mothers in India and also in the USA. Several surrogate mothers face severe morbidities. One surrogate mother told me, "one surrogate mother was very critical and we were all praying that nothing should happen to her and finally she survived but her uterus had to be removed". According to the latest WHO report, nearly five women die every hour in India from complications developed during childbirth, heavy blood loss caused by haemorrhage being a major factor.

While India has shown considerable improvement in literacy levels (65–74% between 2001 and 2011) and school enrolment due to the Sarva Shiksha Abhiyan and the Mid-Day Meal programme in the last 10 years; the Census 2011 household data shows one in ten households still doesn't have even a single literate member (Census of India 2010). Despite a high enrolment rate at primary school, a larger number of girls and boys drop out from school before completing secondary schooling (52 and 53% for girls and boys, respectively) (MHRD 2014). Broken down at different levels, a high percentage of girls are dropping out from schools at primary (27%), elementary (40.6%), and secondary (49.3%) levels (Census of India 2010). Even as there may be little difference in schooling facilities between boys and girls, the opportunities for higher education and employment may be far lesser for women than for men (Sen 2001). In general, education is known to empower women (Navaneetham and Dharmalingam 2002). Higher education among women is positively correlated with access to health care services, enhanced maternal autonomy, child health, health indicators (Vikram et al. 2012). Similarly, higher education of surrogate mothers is known to have an enhanced agency within the surrogacy process (Pande 2010), but most of the surrogate mothers I interviewed had studied only up to middle school level. The doctor in this clinic had said in a TV interview that she has graduates coming as surrogate mothers, but I have never met or even heard of graduate surrogate mothers.

Many of these girls and boys who drop out from schools are married off young. Data shows that women with no education are six times more likely to be married than those with 10 years or more of education. According to the NFHS (National Family Health Survey) data, 47% of the girls are married before the legal age

(18 years) leading to early childbearing (IIPS and Macro International 2007). One in six (16%) girls in their youngest reproductive age group (15–19 years) begin child bearing (IIPS and Macro International 2007). The average age of marriage in India is 17 years and the average age at childbirth is 19 years. Only 57% of India's population participated in the labour force (Gupta et al. 2014). Gender inequalities in India is evident in the low female labour force participation rate (32%) which, according to records is stagnating (NSSO 2013), and the decreasing sex ratio (0–6 age group) from 962 in 1981 to 914 girls every 1000 boys in 2011 (Census of India 2010). There are several reasons economists have presented, one of which are household inequalities, a perception that women need not work until financially required to in the household, hence, may not be inclined to participate in paid employment. There are also ownership inequalities, wherein women owning properties and assets, such as houses or land, are asymmetrically shared. Although legal reforms changed the regulations for property inheritance, ownership continues to be weighed in favour of male children and this may be closely linked with the concept of dowry. Even in the southern part of India, which boasts of socio-economic development, the value of women in family and broader social context has actually regressed in the past 1000 years, has degenerated encouraging the practice of dowry, transpiring into violence against women in India (Mukund 1999: EPW; Rastogi and Therly 2006). There are limitations in the dowry prohibition laws as well as a social inclination and sanction towards maintaining patriarchy (Khanna 2015; Nithya 2013). Women are burdened into choosing to become surrogate mothers to pay off sister-in-law's dowry loans or to save money for their daughter's dowry (RTD 2015). Some scholars may argue that women receive respect and agency in being a surrogate mother. Isn't this a respect that any woman should rightfully deserve as a citizen of India and as a member of a family, community? Instead, women are forced to sell their body parts to maintain one of the most degenerating social practices in the country which is bride-price. Women say, I am doing surrogacy "for the sake of my family", "for the sake of my children", "for the sake of my husband", "my mother-in-law is happy", despite "facing exploitation", "separation from her family", "bodily pains" "psychological effects", "pangs of separation" (Pande 2010; Saravanan 2015). "I would never have taken this step if not for poverty" (refer Chaps. 5 and 6). I have not heard any surrogate mother saying that she is pleased when involved in surrogacy or looking forward to coming back to the surrogate home for another surrogacy. The surrogate mother's husbands invariably leave their jobs and sit at home in the pretext of looking after the children and the women repeat surrogacy until their familial financial goal is fulfilled (see Chap. 5).

Natality inequalities that prefer boys over girls is becoming easily available with modern medical technologies. One such technology increasingly being used for weeding out children in utero for (un)desired characteristics is selective abortions through Prenatal Genetic Diagnosis (PND). Its commercialisation globally is evident in the increasing number of private agencies offering cross-border regional services in Asia, Europe, Middle-East, South America and Northern Africa, USA being one of the most popular destination countries. The global non-invasive prenatal testing (NIPT) market, which was valued at US$0.22 billion in 2012, is estimated to grow up to $3.62 billion in 2019 (Allyse et al. 2015). Although these technologies have been used for treatment of medical ailments found in utero, it is largely being used to decide whether or not to continue with pregnancies (Kahn et al. 2000). Sex selective abortion is a common practice in certain parts of the world accounting for 44 million missing girls in China, 37 million in India, and a total of more than 100 million worldwide (Sen 1992). A 2011 study by a University of Toronto researcher estimated that there have been anywhere from 3.1 to 6 million abortions of female fetuses in India in just the last decade, mostly among wealthy, educated Indians (other estimates suggest the number could be as high as 12 million). In fact, the ratio of girls to boys aged 0–6 is the lowest it has been since the country began recording this data in 1961 (Census of India 2010). In some regions in India where selective abortion is also widespread, the shortage of women has become so acute it has led to a burgeoning bride-trafficking industry in which female children are abducted from one village to be sold off through marriage in another. Selective abortions are not merely individual decisions but deeply embedded within familial, societal, welfare regimes, medical, and legal contexts affecting and being influenced by life course planning, sex preference, and perception of disability reflecting broader social prejudices. This phenomenon too is not confined to certain areas of the world. For instance, sex preference is not merely a problem in Asia or among Asian communities living abroad (SBS 2015). Skewed sex ratios have recently been reported also in the South-East European countries of Albania, Kosovo, Macedonia, and Montenegro, where there seems to be a re-emergence of the preference for a son (Guilmoto 2010). Cross-border movement to access reproductive technologies is known Western European countries where this practice is banned to more permissive countries in Europe. Hence, advancing reproductive technology is a global phenomenon with growing access to cross-border reproductive care and no nation can remain in complete isolation. Many documentaries which have interviewed intended parents in anticipation of the birth of their children through surrogacy in India have declared the sex of the child, despite the fact that it is illegal in India (Journeyman Pictures 2014). In the backdrop of this, commercial surrogacy in India was steadily growing also reinforcing some of the existing inequalities and social prejudices. The growing demand for surrogacy in the circumstances of ineffective governance and existing inequalities made India an ideal environment for global injustice in the process of commercial surrogacy.

Inequalities in the Indian Surrogacy Bazaar

Surrogate mothers in India have comparatively lesser rights over the child compared to the more developed countries in the Global north (UK, USA, Canada). Surrogate mothers in the USA are provided with social support group, insurance for multiple pregnancies, maternity benefits, life insurance, psychological support, compensation for all expenses and loss of employment and representation by an attorney among many other rights and benefits, while none of these are provided to surrogate mothers in India. Both surrogate mothers and intended parents may choose each other and the kind of contract they sign (open or closed), while in India the pattern is chosen and imposed upon both parties by the IVF clinic (Blyth 1994). In India, intended parents participated in selecting surrogate mothers but not vice versa. In the USA, the surrogate mothers are covered with benefits including a grace period post birth to relinquish the baby, a clause with options to withdraw pre-birth and the options for choosing an ongoing relationship with the child and the intended parents (Busby and Vun 2010). Instead, the surrogate mothers in India are exploited by not being provided with a copy of their contract, hence, limiting the possibility of their ever registering a case in the court of law (Saravanan 2013; Puricelli 2014). They are not covered under any insurance and the medical support too culminates with the handing over of the baby. They have to sign off all rights over the child when entering into the contract and have to take the complete responsibility in case of any complications that may arise during pregnancy or childbirth. They are not compensated for loss of employment. They are not given any remuneration for an unexpected miscarriage even this happens as a result of a medically induced selective abortion in utero (Saravanan 2013). Neither the surrogate mothers nor the intended parents in India have the rights to choose the kind of contract, relinquishment, or their preferred mutual relationship.

While in India the primary criterion for selecting, surrogate mothers is poverty, in America, such cases are supposedly screened out (Saravanan 2013). Surrogate mothers in India are detained in hostel-like surrogate homes from the time the embryo is transferred until delivery. They are also expected to care for the baby after birth. They are over-fed and restricted in movements and meetings with their family during their stay in the home. One of the primary human rights violations is their inability to participate in public life and meet non-reproductive aspirations. These practices in India are a violation of basic human rights, dignity, and freedom, as stated in Articles 1,[2] 2,[3] 9,[4] and 14[5] of the Universal Declaration of Human Rights and The Universal Declaration on Bioethics and Human Rights 2005[6]

[2]The Universal Declaration of Human Rights, Article 1 states, "All human beings are born free and equal in dignity and rights. They are endowed with reason and conscience and should act towards one another in a spirit of brotherhood" (UNESCO 2006).

[3]The Universal Declaration of Human Rights, Article 2, states, "Everyone is entitled to all the rights and freedoms set forth in this Declaration, without distinction of any kind, such as race, colour, sex, language, religion, political or other opinion, national or social origin, property, birth or other status" (UNESCO 2006).

[4]The Universal Declaration of Human Rights, Article 9 states, "No one shall be subjected to arbitrary arrest, detention or exile" (UNESCO 2006).

[5]The Universal Declaration of Human Rights, Article 14 states, "Everyone has the right to freedom of movement and residence within the borders of each state".

[6]The Universal Declaration on Bioethics and Human Rights 2005 recognizes that technological advancements in medical science should be ethically sound, giving "due respect to the dignity of the human person and universal respect for, and observance of, human rights and fundamental freedoms" (UNESCO 2006: 3).

(UNESCO 2006; UN 1948). This is not to say that there is absolutely no gender bias in USA, Canada, or the UK, but surrogate mothers are comparatively more medically and legally supported than women in India.

I have published some of my study results which revealed that women in India have no legal or psychological support, they receive a comparatively lesser share of the total surrogacy costs, submit to unfair payment patterns, with no additional payment for miscarriage, are not safeguarded with medical/life insurance, and some clinics make it mandatory that they remain in surrogate homes away from their families (Saravanan 2013, 2015). Surrogate mothers have to sign off all rights over the child while entering into the contract, according to the ART (Assisted Reproductive Technology) Bill 2008. Although legally only three embryos are allowed to be implanted into the surrogate mother's womb in India, up to five embryos are known to be implanted with a high likelihood of multiple pregnancies. In case of these multiple pregnancies (triplets), the doctor suggests 'selective reduction' of one or more fetuses in utero. This procedure may also result in miscarriage and/or preterm labour or infection. The surrogate mother's opinion on the number of embryos to be implanted is not taken nor does she have any decision making power about selective in utero abortions. The surrogate mothers have to sign off all rights on medical interventions. The other inequalities that would not be found in other developed countries but in India is the mandatory rule imposed on women to stay in dormitory homes also called the surrogate homes throughout their pregnancy. I have written more about this in Chap. 3, the psychological impact of such seclusion is not studied. One surrogate mother asked permission to go home because her mother was unwell and the nurse had asked her, "So when are you coming back? You'll have to come back in two or three days as your mother feels better" (SAMA and Sharma 2010). "They don't leave us even for five minutes, they don't trust, what if we run off home?" (SAMA and Sharma 2010). One surrogate mother in a home in Mumbai said, "My family comes to meet me but it's been three months and I have not gone home" (SAMA and Sharma 2010). The broker plays a role in policing the surrogate mother, her relationship with the child, using persuasion, coercion for money and threat of lawsuit to weaken and destroy whatever maternal love she may develop for her child. To Dr. Nayna Patel, her fertility centre is a one-stop shop "right from the embryo transfer until the baby is handed over, we offer everything in one place" (Time 2015).

Surrogate Homes

The restrictions imposed on surrogate mothers in India to stay in surrogate homes violate their basic human rights to participate in public life and meet non-reproductive aspirations during pregnancy, which is contrary to liberating. In such a facility, there is an element of medical and social scrutiny for food, diet, sexual behavior, mobility, and even daily household work. The ethical concern that arises from the making of such homes mandatory is that women confined to these homes are unable to participate fully not only in their personal lives or also in public life; they are restricted in participating in their non-reproductive aspirations such as education, occupation, and social functions; hence, the "institutional surrogate mother" becomes treated as the "means to an end". Some scholars highlight that living in such homes facilitates networking and bonding (Pande 2010; Rudrappa 2012). However, too much focus on non-progressive subjectivity and micro-level autonomies that women exercise within the process of surrogacy mystify the larger picture of structural inequalities and injustice. Some women themselves become consumers or surrogacy agents post surrogacy and also exploiters coercing women in desperate conditions into this process. My study revealed that women who are involved in drug trials and gamete donation and even prostitution are easy recruits into the process of surrogacy as they are already involved in the process of body sale in the bio-markets. Families involved in these activities are aware that women's body parts give more monetary returns and surrogacy is the most remunerative and lucrative option for quick money making. Infertile poor women are excluded from this biomaterial markets although some are coerced into oocyte sale. This is a chance for fertile young women to prove their dedication towards the family and indeed most women are willing participants to surrogacy. Those who are first introduced into this market through surrogacy gradually also get involved in gamete donation and drug trials. Drug trial centres are primary locations where recruitment of potential surrogate mothers takes place. Surrogate agents comb these places and also low-income localities to recruit women in dire circumstances.

Hence, the position of the consumer and service providers is stratified yet interchangeable in this neocolonial reproductive market. The consumers of the reproductive biomaterial in the surrogacy markets include medical practitioners; agents; and hotels, commuting service, accommodation, and food (tiffin box) providers; as well as other associated small business activities. However, the biggest profiteer among all these are the owners of the fertility clinic. They are profit-making institutions that aim at enhancing the bioavailability of surrogate mothers and optimizing their medical services in comparison to other service providers in a competitive market. The services they provided were largely based on the preferences of the intended parents.

Normalization of Surrogacy as Work

Both Pande (2010) and Rudrappa (2012) consider surrogacy as a work, an extension of employment such as factory work. Rudrappa claims that "given their employment options and their relative dispossession, they believed that Bangalore's reproduction industry afforded them greater control over their emotional, financial, and sexual lives. In comparison to garment work, surrogacy was easy" (Rudrappa 2012: 27). Medical practitioners in India draw comparisons between surrogacy and housemaid work (NDTV). It is well known that garment factory work is close to slavery, it is legalized, and there is no attempt to improve their working conditions. The inhuman working conditions is ongoing unabated and the supply is exported and governments import such products. People continue to buy cheap clothes sold in the markets abroad and the inhuman market also goes on unabated in Bangladesh, Sri Lanka, and India (Gunatilaka 2010; Stanwick and Stanwick 2015; Saha 2014). These studies found high incidences of injuries on the job, low wage rates, misuse of gender discrimination, substandard air quality, and lack of safety factors. The surrogacy industry in India is largely an inequality-based, demand-driven market feeding into the needs of affluent countries of the global North and those within India. Implementation has been weak in this sector owing to the ineffective governance along with power of the medical fraternity. Even when surrogacy was permitted in India, tribal women were being still trafficked into surrogacy. Indian women were being transported to Nepal for surrogacy. Some surrogate mothers have died, others have faced serious morbidities, but these cases have been hushed up and legally the concerned clinics, doctors, and agencies were convicted because such operations were not declared illegal.

A common question raised is that banning surrogacy in India would mean the practice would go underground in worse forms. Experience shows exactly the opposite. Sex determination and selective abortions was rampant in India when there was no law against this practice. However, the prohibition of sex determination and sex selective abortions in India has proved to have had a positive impact over a period of time. A recent analysis of the effect of the Pre-Conception and Pre-Natal Diagnostic Techniques (PCPNDT) Act, 2011 using a treatment–effect analysis framework concluded that the law implementation has had a significant impact in preventing an extreme worsening of the gender imbalance. A possible absence of the law would have led to at least 106,000 fewer girl children in India (Nandi and Deolalikar 2013). Despite strong criticism from liberals for being radical, feminists in India have strongly opposed sex selective abortions as a form of "femicide" and "violence against women" (Patel 1989; George 2006; Sharma 2001). The Supreme Court of India imposed strict regulations on the use and sale of ultrasound machines since 2001, if medical practitioners are found guilty of this offence they are required to pay a fine and have their medical license withdrawn. Twenty years of implementation of the Prenatal Diagnostic Technology (PNDT) Act 2004 has revealed that the law has been largely effective in controlling further elimination of girls. It is only the most powerful doctors who are confident of evading the law and continue conducting sex determination (Saravanan 2017). Raymond (1993) observes that legal prohibition can be a useful tool in controlling certain human rights violations.

Surrogacy is justified as a better work option than the garment export markets (Rudrappa 2012). The working conditions of surrogate mothers and housemaids are similar, both unprotected, informal, with no life/medical insurance, over-worked and underpaid experiencing confined labour. As many surrogate mothers are working either as housemaids or garment factories and both these occupations are exploited, these women dominated informal sectors should be improved. Garment factory workers and housemaids should be brought into formalized markets with improved working conditions, increased remuneration, and be protected by social security such as retirement fund. This would be a long-term opportunity for women rather than to encourage surrogacy practice which has detrimental physical and psychological impacts on women. Formalization and enhanced social security of informal labour will make changes in the lives of women seeking to be surrogate mothers. The word "*kaam*", often used by surrogate mothers in India is grossly misinterpreted as surrogate mothers describing surrogacy as "work". When asked about their motivations they tend to say "*hum achha kaam kar rahe hain*". A virtual translation of the word "kaam" is "deed" as well as "work", while what surrogate mothers mean by saying "achha kaam" is "good deed" or a "noble service", this has to do with their altruistic motivations rather than their reference to surrogacy as work. In the Indian cultural context, prostitution or infidelity is considered bad "*karma*" and this is what they are clarifying by saying that the IVF process does not include sleeping with anybody and also in the end gives a child to a couple, hence, it is "achha kaam". They are not saying that surrogacy should be a job option that all women should consider as a career, an employment that the government should promote for all women, nor are they saying that they would ever suggest this as an option for their daughters in the future. One surrogate mother in my study said, "I am doing this (surrogacy) for my children. I am going through this agony so that my children have a better future, so that they can go to a good school, have a good education and need not do anything close to this in their lives and can manage to earn enough by doing other kinds of work." Hence this is not a job option they would suggest for women who are not in dire need of money. Pande (2010) interprets the surrogate mother's use of the word "*majboori*" as "a necessity" and that they were downplaying their choice to become a surrogate mother. I vary in my understanding, by the word "majboori", surrogate mothers meant "helplessness" of poverty that motivated them into surrogacy. By saying "majboori", surrogate women were expressing their vulnerability.

Normalization of classist prejudice is also evident in various other forms of division of labour. It is normal for the affluent around the globe to outsource unskilled jobs to the socio-economically disadvantaged people. Housework or childcare that is outsourced to maids and nannies is a widely accepted practice (Das et al. 2015). In the same light, outsourcing of surrogacy to poor women and their exploitation in India is not considered any different. These patterns are hence, easily translated into other forms of reproductive labour such as surrogacy. A practicing IVF doctor in Mumbai, Aniruddha Malpani says, "Exploitation is normal in India; doctors exploit their patients; lawyers exploit their clients, everybody exploits every other person, so what?" (Saravanan 2010). Although both Pande (2014) and Rudrappa (2012) agree that surrogacy in India is exploitative, they perceive that women are empowered in the process. Surrogate mothers in all documentary films, research, and in my study say they would never have taken such a big step if not for abject poverty (Haimowitz and Sinha 2010). So, given a choice, these women would never want to be a surrogate and in case they had a better payment and any sort of security cover in their other jobs they would not have opted for surrogacy. Dr. Nayna Patel justifies, "It's like a maidservant saying that I would never want my daughter to become a maidservant, or a labourer saying I don't want my son to be a labourer." (James 2013). Similarly, some academics claim, the risks in surrogacy are similar to any other kind of work, all forms of the normalizing language used in this industry. Some of the key export industries selling products in the global North (garments, leather goods, textiles, shoes, carpets, and electronics) are based on extremely exploitative labour conditions in Asia and Africa. Just like surrogacy, this industry too is largely governed by the powerful producers and retailers, while the bottom of the chain typically benefit least from this arrangement. Yet Dr. Nayana Patel perceives it as an agency and insists, "They (the surrogate mothers) make a voluntary decision to opt for this and earn some money. How can you deny them the right? The argument that they are exploited is not true" (NDTV 2015).

According to Pande (2010), the cheap, docile, selfless, surrogate mother is not available ready-made but she is trained by the clinic to be so to fit these qualities to suit the global markets. In my study too, the first explanation given to the women is that she will rent her womb and she has nothing to do with the baby as the gametes belong to someone else, she only has to carry the baby and hand it over for a remuneration. Inertness and submissiveness are the traits sought by IVF clinics in a surrogate mother. Women were politely rejected on medical grounds when they are observed to have aggressive characteristics. Surrogate mothers had to submit to medical procedures, a predetermined compensation, stipulated accommodation arrangements, and were expected to attend all appointments and behave in a certain manner within the surrogate homes. It was only when I met them after relinquishment or away from the clinic atmosphere that they shared their experience in entirety with me. The medical practitioners expected surrogate mothers to be healthy, clean, understanding, and conscious of their responsibilities. Consent from the surrogate mothers' family members, especially the husband, is of importance to them as they don't want family members coming and fighting with them about recruiting women in their family as surrogate mothers. The ART Bill in India also required surrogate mothers' husbands to sign, stating their consent for the surrogacy. This reveals the social position of women who are not even included in decisions regarding their own health care.

Amrita Pande (2010) says surrogate mothers are workers, not wombs, based on the phrase "dirty work" which is used by surrogate mothers and by women defying this stigmatization. However, Pande's view lies in contrast to her book titled *Wombs in Labor* which alienates women's labour from womb (Pande 2014). Coming to women's agency, Pande (2010) rightly points out that whatever little resistance there is, it is not collective but individual. Surrogacy is kept at the familial level and any resistance is not intended to overthrow power, emancipate, or change the system. Unlike Pande's (2010) findings where most surrogate mothers kept surrogacy a secret from their community and family, most of the surrogate mothers in my study had informed their relatives at some time during the pregnancy or soon after. Some women in my study entered surrogate home along with other women in the extended family.

Regarding their relationships with the child(ren), in the documentary made by SAMA, surrogate mothers said, “The child is my *godh*[7] (womb), so it’s mine, but I gave it in your hand. I will only pray to God to protect them” (SAMA and Sharma 2010). But Dr. Nayna Patel’s understanding is limited to the intended parent’s point of view “who gives one’s own child to another person to care for 9 months?” (RT Documentary 2015). The surrogate mother further added, “Wherever my child is, protect it, no harm should befall my child. Whenever I pray in front of God, I tell God I have three children, protect all the three children; I have never met or seen, but if I bless from my heart, my child will receive the blessings.” They expressed the intent to be able to keep in contact and see the baby. Regarding the intended parents, whom they considered a saviour, surrogate mothers said, “(The parents) whose child it was, they had helped me a lot. They offered to fulfil all my needs, need for a house, money.” The sense of hierarchy was very much present, “I met the intended parents only once and they told me to listen to bhajans for two hours every day. They wanted a good influence on the child. At least my intended parents were Indian so they know, but the foreigners cannot communicate in the same language so the surrogate mothers don’t understand” (SAMA and Sharma 2010).

It is presumed that women in India earn much more money in surrogacy than they ever could in their present jobs; they also receive unimaginable support for housing and healthcare during their pregnancy (Dale-Moore 2015). Such arguments infer that women are better off being involved in surrogacy than being unemployed or working in lesser salaried positions and that women’s health has improved just by practicing in surrogacy.

Work-up for the Payment

One of the arguments made is that women are paid for their reproductive labour and not for the child (Robertson 1994). The liberals support commercial surrogacy suggesting that it is not the child or fetus but the woman’s gestational services that is compensated. Indeed, they are paid for the reproductive labour leading to the height, weight, ability, and sex of the child. One medical practitioner said, “They know their payment will be made only after the check-up. If the check-up is not good, their payment will be delayed. It’s not only an incentive for them but they know they have to work up otherwise they will not be paid” (SAMA and Sharma 2010). “We get Rs. 10,000 extra, if the baby is 3 kg and 100 g, and if the baby weighs more than that then some more extra money is given. They weigh us every second week.” To which another surrogate mother joked, “When we go for

[7]The virtual translation of the word ‘godh’ would be ‘lap’, but in this context surrogate mothers mean ‘womb’.

weighing we should wear some stones around our neck". Accordingly, they said, "There is no dearth of food, whatever you want they give, eggs, a packet full of vitamin biscuits has to be eaten. But we can't eat so much. We don't get sleep, we are awake all night, we only eat and sleep. One can sleep only if one gets tired, if we are not tired how can we sleep? When we had our children, we ate and then worked in the house so we never felt restless. But here we just eat and sit around so we can't sleep nor does our food get digested." Another surrogate mother said, "I was 43 kg, here the Madam increased my weight to 70 kg, I became healthy after I came here" (SAMA and Sharma 2010).

Surrogacy as a Hobby, a Frivolous Activity of the Rich

That surrogacy is also frivolously motivated is not a supposition any more. In my study with few (13) participants, one surrogate mother, Shama, in a clinic from Ahmedabad, told me her Indian intended mother was very scared of pregnancy and delivery, hence she hired her. Shama explained to me saying, "He (the intended father) is a big businessman, they earn a lot of money so they don't have time for this (pregnancy)". "Many first try the treatment on themselves and when it doesn't work they come for surrogacy, but this is not right." It was atrocious that the intended mother asked Shama, "Don't you feel scared, I am scared," to which Shama replied, "I have two children, I am not scared". Shama repeatedly told me about this and laughed. Eventually, she hit head lightly and said "I didn't know there are such people also in this world who are hiring surrogate mothers because they are scared to have a child and I wonder why they selected me. There were so many women in the recruitment folder and why did she have to select me? She sympathized with the man, "what can the husband do, if the lady says 'no', I am sacred". I have written more about Shama in Chap. 5. I would like to reiterate that I am not trivializing the fear of pregnancy that some women may experience but this fear with a combination of geneticization, affluence to outsource this pregnancy, and markets that make such possibilities easily available, and exploitation of poorer women brings the process dangerously close to being inhuman.

As the External Affairs Minister of India, Sushma Swaraj rightly observed Indian cinema celebrities recently found it a convenient way of having children despite having children of their own. She said, "The procedure that started as a necessity has become a hobby of sorts. We have many examples of celebrities who have their own children, still they have gone for a surrogate child." Surrogacy cases such as Shama's, is rarely known to public or media, but when celebrities do the same, people tend to normalize and easily follow their example. Recently a popular Hindi film star, Shahrukh Khan despite having two teenage biological children decided to have a third child through surrogacy. Similarly, Aamir Khan has biological teenage children but went on to have another child through surrogacy. While the rich and famous fuel a rise in surrogacy practice, child adoption rates have dropped in India.

IVF experts and medical organizations such as the ICMR believe that the surrogacy option has made the adoption idea less popular (Sharma 2016). This is evident in the decreasing number of adoptions within India, from 5,964 adoptions in 2010–11 to 3,210 in 2016–17 (CARA 2017). The ideological trend towards "I have the right to have a baby and with my own genome" can be dangerous especially given the extent of commodification in this industry.

Film stars like Shahrukh Khan, Aamir Khan, Tusshar Kapoor, Sohail Khan had one boy each using surrogacy. Is this just a coincidence or was sex selective technology used in the surrogacy process, given the strong patriarchal presence in India? People have started using IVF and surrogacy just to be able to choose the sex and other characteristics of the child. Take other examples from the film fraternity; recently, another prominent celebrity, Karan Johar was in news in India because he had used surrogacy to have twins; one girl and one boy child, a perfect balanced family. The elite and influential may indulge in sex selection of embryos and gamete selection before IVF, all to suit their fancies and needs.

Conclusion

Corea (1985) strongly objected to surrogate motherhood as it creates divisions among women referring to this phenomenon as "segmented reproduction" that divides women into childbegetters, childbearers, and childrearers. She critiques this segmentation as if it were a mode of production of genetically superior women begetting embryos, strong-bodied women bearing the babies to term, and sweet-tempered women rearing the infants to adulthood. Colen's (1995) concept on stratified reproduction is similar to Corea's reference as it examines reproductive labour of bearing, rearing, and socializing children that may be differentially experienced, valued, and rewarded according to inequalities of access to material and social resources structured by hierarchies of class, race, ethnicity, gender, place, and migration status differences, aspects of which promote or interfere with socio-economic and political status. In today's context of surrogacy bazaars, this concept is not fully adequate because it does not include the role of medical practitioners, or surrogate agents, and it excludes the body politic. Subsequent chapters examine the role of intended parents (the childbegetters), surrogate mothers (childbearers), and childrearers (the surrogate mothers for a short while and the intended parents) interviewed in this study in 2 IVF centres in two places in western India. This chapter already revealed that these participants had differing experiences, were valued differently, and were rewarded differently for the reproductive labour based on socio-economic hierarchies. This will be elaborated in the following chapter. The historical patterns of exploitation and discrimination, hence, postcolonialism will be discussed along with the experience of intended parents in Chap. 6.

References

Allyse, M., M.A. Minear, E. Berson, S. Sridhar, M. Rote, A. Hung, and S. Chandrasekharan. 2015. Non-invasive prenatal testing: A review of international implementation and challenges. *International Journal of Women's Health* 16 (7): 113–126.

Bhalla, Nita, and Mansi Thapiyal. 2013. *India seeks to regulate its booming 'rent-a-womb' industry*. *Reuters*. http://www.reuters.com/article/us-india-surrogates-idUSBRE98T07F20130930. Accessed 14 Aug 2017.

Blyth, E. 1994. I wanted to be interesting. I wanted to be able to say 'I've done something interesting with my life': Interviews with Surrogate Mothers in Britain. *Journal of Reproductive and Infant Psychology* 12 (3): 189–198.

Busby, K., and D. Vun. 2010. Revisiting the Handmaid's tale: Feminist theory meets empirical research on surrogate mothers. *Canadian Journal of Family Law* 26 (1): 13–93.

CARA. 2017. *Adoption statistics*. Central Adoption Resource Authority, Ministry of Women & Child Development, Government of India. http://cara.nic.in/resource/adoption_Stattistics.html. Accessed 14 Aug 2017.

Census of India. 2010. *Population data*. India: Registrar General, Government of New Delhi.

Colen, S. 1995. 'Like a mother to them': Stratified reproduction and West Indian childcare workers and employers in New York. In *Conceiving the new world order: The global politics of reproduction*, ed. Faye Ginsburg, and Rayna Rapp, 78–102. Berkeley: University of California Press.

Corea, Gena. 1985. *The mother machine*. Reproductive technologies from artificial insemination to artificial wombs. New York: Harper and Row.

Dale-Moore, Rosa. 2015. A Consideration of International Surrogacy Under Act-Utilitarian Ethics. *Sound Decisions: An Undergraduate Bioethics* 1 (1). Available at: https://soundideas.pugetsound.edu/sounddecisions/vol1/iss1/1.

Das, Sonali, Sonali Jain-Chandra, Kalpana Kochhar, and Naresh Kumar. 2015. *Women workers in India: Why so few among so many?* International Monetary Fund (IMF) Working Paper, WP/15/55, 1–31.

Desai, Kishwar. 2012. India's surrogate mothers are risking their lives: They urgently need protection. *The Guardian*. https://www.theguardian.com/commentisfree/2012/jun/05/india-surrogates-impoverished-die. Accessed 15 Aug 2017.

Fertility Consultants Canada. 2014. What is surrogacy and how much surrogate mother cost in Canada? *Fertility Consultants Canada*. https://fertilityconsultantscanada.wordpress.com/2014/04/11/what-is-surrogacy-and-how-muchsurrogate-mother-cost-in-canada/. Accessed 9 Feb 2015.

Frontline. 2016. *Surrogacy a hot topic*. Interview: Dr. Sowmya Swaminathan. *Frontline India*. http://www.frontline.in/social-issues/surrogacy-a-hot-topic/article9140558.ece. Accessed 15 Aug 2017.

George, S.M. 2006. Millions of missing girls: From fetal sexing to high technology sex selection in India. *Prenatal Diagnosis* 26: 604–609.

Guilmoto, C.Z. 2010. Longer-term disruptions to demographic structures in China and India resulting from skewed sex ratios at birth. *Asian Population Studies* 6: 3–24.

Gunatilaka, R. 2010. *Sri Lanka's working poor*. Colombo: ILO.

Gupta, R., S. Sankhe, R. Dobbs, J. Woetzel, A. Madgavkar, and A. Hasyagar. 2014. *From poverty to empowerment: India's imperative for jobs, growth and effective basic services*. Mumbai: McKinsey Global Institute.

Haimowitz, Rebecca, and Vaishali Sinha. 2010. Made in India. DVD.

ICMR. 2015. List of enrolled assisted reproductive technology (ART) clinics under national registry of ART clinics and banks in India. *Indian Council of Medical Research*. http://www.icmr.nic.in/icmrnews/art/List%20of%20Enrolled%20ART%20Clinis-5%203%202013.pdf. Accessed 6 Apr 2015.

IIPS and Macro International. 2007. *National family health survey 3 (2005–2006) India*, vol. I. Mumbai: National Fact Sheet. International Institute for Population Sciences (IIPS).

James, Mac. 2013. *Surrogates have positive experience. YouTube*. www.youtube.com/watch?v=fXnqmKUu6rY. Accessed 14 Aug 2017.

Journeyman Pictures. 2014. Commercial surrogacy exploiting women of the developing world. *Bittu TV*. http://channel-v.bittutv.com/?watch=Rj3EodH7lcY. Accessed 14 Aug 2017.

Kahn, J.G., B.J. Becker, L. MacIsaa, J.K. Amory, J. Neuhaus, I. Olkin, and M.D. Creinin. 2000. The efficacy of medical abortion: A meta-analysis. *Contraception* 61 (1): 29–40.

Kannan, S. 2009. *Regulators eye India's surrogacy sector. BBC World*. http://news.bbc.co.uk/2/hi/business/7935768.htm. Accessed 14 Aug 2017.

Khanna, Leela. 2015. The state and domestic violence: The limitations of India's dowry prohibition laws. *Senior Projects Spring* Paper 150.

MHRD. 2014. Department of School Education & Literacy Department of Higher Education Ministry of Human Resource Development Government of India, Annual Report. Government of India, New Delhi.

Mukund, Kanakalatha. 1999. Women's property rights in South India: A review. *The Economic and Political Weekly* 34 (22): 1352–1358.

Nandi, Arindam, and Anil B. Deolalikar. 2013. Does a legal ban on sex-selective abortions improve child sex ratios? Evidence from a policy change in India. *Journal of Development Economics* 103: 216–228.

Navaneetham, Kannan, and Arunachalam Dharmalingam. 2002. Utilization of maternal health care services in Southern India. *Social Science and Medicine* 55 (10): 1849–1869.

NDTV. 2015. We the people—The surrogacy debate; of motherhood or money? *YouTube*. www.youtube.com/watch?v=7OlBHVKZzjg. Accessed 14 Aug 2017.

NDTV. 2016. Women selling bodies to earn livelihood: Minister Anupriya Patel on surrogacy bill. *New Delhi TV*. http://www.ndtv.com/india-news/women-selling-bodies-to-earn-livelihood-minister-on-surrogacy-bill-1453387. Accessed 15 Aug 2017.

NFHS. 2007. *National Family Health Survey 3 (2005–2006) India*: vol. I. Mumbai: National Fact Sheet. International Institute for Population Sciences (IIPS).

Nithya, N.R. 2013. Institutionalization of dowry in India: Social custom or modern Malaise? *International Journal of Science and Research (IJSR)* 2 (9): 46–50.

NSSO. 2013. *Key indicators of employment and unemployment in India*. New Delhi: The National Sample Survey Office (NSSO), Ministry of Statistics and Programme Implementation.

Pande, A. 2010. Commercial surrogacy in India: Manufacturing a perfect mother-worker. *Signs* 35: 969–992.

Pande, A. 2014. *Wombs in labor: Transnational commercial surrogacy in India*. New York: Columbia University Press.

Patel, Vibhuti. 1989. Sex-determination and sex pre-selection tests in India: Recent techniques in femicide. *Reproductive and Genetic Engineering* 2: 111–119.

Perappadan, B.S. 2014. Activists call for stringent regulations for surrogacy. *The Hindu*. http://www.thehindu.com/todays-paper/tp-national/tp-otherstates/activists-call-for-stringent-regulations- for-surrogacy/article6348214.ece. Accessed 15 Aug 2017.

Puricelli, M. 2014. Conference: A policy dialogue about surrogacy in India. *India: Gender Matters*. http://csrindia.org/blog/2014/09/18/surrogacy-in-india. Sep 18.

Qadeer, I. 2009. Social and ethical basis of legislation on surrogacy: Need for debate. *Indian Journal of Medical Ethics* 6 (1): 28–31.

Rastogi, M., and P. Therly. 2006. Dowry and its link to violence against women in India: Feminist psychological perspectives. *Trauma Violence and Abuse* 7 (1): 66–77.

Raymond, J.G. 1993. *Women as wombs: Reproductive technologies and the battle over women's freedom*. San Francisco: Harper.

Robertson, John A. 1994. *Children of choice: Freedom and the new reproductive technologies*. Princeton: Princeton University Press.

Russia Today Documentary. 2015. Wombs for rent in India. *YouTube*. https://www.youtube.com/watch?v=PSXZSdMmRdg. Accessed 14 Aug 2017.

Rudrappa, Sharmila. 2012. India's reproductive assembly line. *Contexts* 11 (2): 22–27.

SAMA, and S. Sharma. 2010. *Can we see the baby bump*. India: Surfilms.

Saha, Sudeshna. 2014. Women employees in garment industries—A case study conducted in selected garment industries of Peenya Industrial Area, Bangalore. *International Journal of Management Research and business Strategy* 3 (3): 128–140.

Saravanan, S. 2010. Transnational surrogacy and objectification of gestational mothers. *Economic and Political Weekly* 45: 26–29.

Saravanan, S. 2013. An ethnomethodological approach to examine exploitation in the context of capacity, trust and experience of commercial surrogacy in India. *Philosophy, Ethics and Humanities in Medicine* 8: 10.

Saravanan, S. 2015. Global justice, capabilities approach and commercial surrogacy in India. *Medicine, Health Care and Philosophy* 18: 295–307.

Saravanan, S. 2017. When doctors rock the cradle. *Governance Now* 8: 56–58.

Sarojini, N., V. Marwah, and A. Shenoi. 2011. Globalisation of birth markets: A case study of assisted reproductive technologies in India. *Global Health* 7: 27.

SBS. 2015. Could gender-selective abortions be happening in Australia? *Special Broadcasting Service*, Australia. http://www.sbs.com.au/news/article/2015/08/17/could-gender-selective-abortions- be-happening-australia. Accessed 14 Aug 2017.

Sen, A. 1992. *Inequality re-examined*. Oxford: Clarendon Press.

Sen, A. 2001. Many faces of gender inequality. *Frontline* 18 (22): 35–39.

Sharma, Dinesh C. 2001. Indian government must enforce ban on gender diagnostic test. *The Lancet* 357: 1599.

Sharma, Neetu Chandra. 2016. *'It's what the rich and famous do': Indian celebrities fuel a rise in surrogate births, while child adoption rate drops*. Britain: Mail Online India.

SRS. 2013. *Maternal mortality ratio (MMR) (per 100000 live births). Sample Registration Survey*. New Delhi, India: Registrar General and Census Commission.

Stanwick, Peter, and Sarah Stanwick. 2015. The garment industry in Bangladesh: A human rights challenge auburn university department of management. *Journal of Business & Economic Policy* 2 (4): 40–44.

The VJ Movement. 2009. Rent-a-womb: Outsourcing Surrogacy in India. *YouTube*. https://www.youtube.com/watch?v=GBHs6OI6xIs&t=1s. Accessed 14 Aug 2017.

Time. 2015. Outsourcing surrogacy. *Red Border Films*. *YouTube*. https://www.youtube.com/watch?v=N9FPiNc6-dI. Accessed 14 Aug 2017.

Today. 2008. More and more couples finding surrogates in India. 28th February. https://www.today.com/news/more-more-couples-finding-surrogates-india-wbna23252624.

UN General Assembly. 1948. Universal declaration of human rights 217 A (III). http://www.refworld.org/docid/3ae6b3712c.html. Accessed 19 May 2016.

UNESCO. 2006. *Universal declaration on bioethics and human rights*. United Nations Educational, Scientific and Cultural Organisation. http://unesdoc.unesco.org/images/0014/001461/146180E.pdf. Accessed 17 May 2016.

Vikram, Kriti, Reeve Vanneman, and Sonalde Desai. 2012. Linkages between maternal education and childhood immunization in India. *Social Science and Medicine* 75 (2): 331–339.

Chapter 5
Surrogacy Biomarkets in India: Stratified Reproduction and Intersectionality

Abstract Feminists have noted the ways in which certain power relations in stratified reproduction empower some people to nurture and reproduce while disempowering others (Colen in Conceiving the new world order: the global politics of reproduction. University of California Press, Berkeley, pp. 78–102, 1995; Ginsburg and Rapp in Conceiving the new world order: the global politics of reproduction. University of California Press, Berkeley, 1995; Rapp in Soc Res 78:693–718, 2011). Transnational surrogacy in India reflects many of these inequities; disparities in gender, race, class, and location place some women's reproductive projects above others (DasGupta and DasGupta in Globalization and transnational surrogacy in India: outsourcing life. Lexington Books, Lanham, 2014; Gupta in Eur J Women Stud 13:23–38, 2006, Gupta in IJFAB: Int J Fem Approach Bioeth 5:25–51, 2012; Pande in Reprod Biomed Online 23:618–625, 2011). This chapter examines the experience of surrogate mothers from a feminist perspective by applying Colen's (Conceiving the new world order: the global politics of reproduction. University of California Press, Berkeley, pp. 78–102, 1995) stratified reproduction as a framework to understand how the reproductive tasks of bearing and rearing children is differentially experienced, valued, and rewarded according to inequalities and whether these inequalities structured by socio-economic and political status lead to reproductive/social injustice and human rights violation. Drawing on my fieldwork, the chapter describes the socio-economic background and motivation of surrogate mothers, the role of the medical practitioners, and the embodied surrogacy experience of surrogate mothers and intended parents with reference to the inequalities raised in the previous chapter.

Keywords Stratified reproduction · Ethonology of surrogacy practice
Pregnant embodiment · Intersectionality of poverty-gender-race

Using ethnomethodology, I have tried to identify the phenomena, as perceived by the actors in a situation, giving importance to their interpretations, by conducting

S. Saravanan, *A Transnational Feminist View of Surrogacy Biomarkets in India*,
https://doi.org/10.1007/978-981-10-6869-0_5

semi-structured interviews with a total of 13 surrogate mothers,[1] four intended parents, and two doctors from two clinics in Western India (one with a surrogate home and one without a surrogate home) from August 2009 and April 2010. Aasha clinic (with surrogate homes) in a small town in Gujarat introduced me to Nargisa, Dimpy, and Sumita and allowed me to enter into the dormitory home. My interviews and in-depth interactions with the surrogate mothers led me to conduct participant observation with five surrogate mothers (Nargisa, Dimpy, Ujwala, Sarala, and Gargi) and two intended parents (Caroline-Canadian and John-German). All the surrogate mothers gave me their contact details so that I could keep in touch with them and meet them outside the clinic. Thereafter, I became involved in participating in the surrogacy process as a translator between the surrogate mothers and the intended parents in the clinic during the nanny care service provided by the surrogate mothers. I became a translator in the process of nanny selection, when intended parents recruited nannies to take back with them to their respective country, as well as a spokesperson for the surrogate mothers when they wanted to make requests to the intended parents. The number of case studies are fewer compared to other studies because the methodology I used was an intensive, time-consuming process. I also adopted a third method; with some of the surrogate mothers whom I interviewed I asked five of the surrogate mothers to write the highlights of their own surrogacy experience. As some of the surrogate mothers couldn't write confidently, I had employed Sarala (an ex-surrogate mother) as my research assistant to write while they spoke. The other clinic in Ahmedabad, a larger city in Gujarat (without surrogate homes), introduced me to Gargi and Shama. Shama did not want to give me her contact details; I later came to know through Gargi that the clinic had given strict instructions not to share with me their contact details and their remuneration. I got to know after one interview with a couple and a brief interview with Gargi that there was a huge discrepancy between what was taken by Dr. Bhargav from the intended parents as surrogate mother's fees and what was actually paid to them. Such a payment discrepancy was also prevalent in Aasha Clinic. Aasha Clinic charged the intended parents excessively under all kinds of pretext—costs incurred for caesarean delivery, dormitory expenses—apart from the bonus money that they charged as nanny expenses, though in practice the surrogate mothers provided such services after the child's birth, including breastfeeding. Dr. Nisha at Aasha Clinic had made it mandatory that all cash transactions between the intended parents and the surrogate mothers would strictly be made through the clinic only. All the quotes given in this chapter are taken from the study that I conducted between 2009 and 2012 and all names are pseudonyms to maintain the confidentiality of participants; wherever I have cited from outside the study I have referenced it accordingly. Sarala was Dimpy's sister-in-law, also a surrogate mother; Gargi is the surrogate mother I interviewed

[1]Nargisa, Dimpy, Ujwala, Sarala, Gargi, Sumita, Uma, Shama, Urmila, Maya, Sheetal, Sapna, Mamta.

and interacted with from the second clinic in Ahmedabad. I interacted with all these mothers intensively and am still in contact with them. I have referred to Gargi's experience in Chap. 6 as it has a significant component of how the intended parent's desires and exploitation directly affected her experience, her sense of value and reward. Four intended parents were interviewed using semi-structured interviews, two each from Europe (John and Mary) and North America (Caroline and Shyamala). The discussions are placed in the context of postcolonialism, as well as examining Colen's (1995) notion of stratified reproduction as mentioned in Chap. 6.

The semi-structured interviews were conducted in Hindi as all surrogate mothers spoke Hindi or a mixture of Hindi and Gujarati. I am well versed in both languages.[2] I recorded the interviews and translated them into English and transcribed for reading and rereading. The process of analysis included: transcription (not only of the literal statements recorded on tape, but also of the non-verbal gestures made during the conversations), bracketing and phenomenological reduction, listening to the tape and reading the transcription repeatedly to provide the context of smaller units of meaning for deriving emerging themes and delineating units of meaning relevant to the research question and the methodological framework. These were then clustered into themes of emotions and experiences, such as knowledge, trust, coercion, fear, feelings of mistrust and dependency reflecting on the differences in embodiment of experience, value, and reward, based on inequalities. The topics covered in the semi-structured interviews included socio-economic backgrounds, motivations, experiences of rules, living in the surrogate home, bonding, financial dealings, relinquishment, and post-relinquishment. Diary notes were maintained for the participant observation. Written consent was obtained from the clinic' s principal medical practitioner, Dr. Nisha, through email before my arrival in India. Written consent was also secured from the research participants, using consent forms. The consent form was translated into Hindi and Gujarati by local professional translators and was written in simple understandable language. All surrogate mothers could read and were more than willing to participate. There was no formal process of ethical clearance at the University of Heidelberg, Germany, at the time of the research but this has been introduced recently.[3]

---

[2]I was brought up in Mumbai and lived in a Gujarati colony; hence, I had learnt English and Hindi in school and Gujarati from my neighbours.

[3]However, for this particular research, a research team from South Asia Institute and Karl Jaspers Centre Heidelberg, as well as a few invited international research experts, generally discussed and screened the proposed study, both academically, structurally and ethically, at the workshop, "Making India a Global Healthcare Destination: Historical and Anthropological Enquiries on Cross-border Healthcare". This was co-organized by the Cluster of Excellence "Asia and Europe", Heidelberg, and the French Institute of Pondicherry, held in Heidelberg, 14–15 June 2009.

Socio-economic Background and Motivation of Surrogate Mothers

All the surrogate mothers I interviewed were facing severe economic difficulties at home; two women had an ill family member each, one had a child with severe disabilities and the other's husband had a heart disease which needed immediate medical treatment. There were another three women who found it difficult to make ends meet due to their small earnings. One woman's husband was an alcoholic addict and spent most of his income on this habit. Four women wanted to save money for their children's education. Two women wanted money to rebuild their "*kachha*" (unstable, temporarily built) house. One wanted to buy a house, since most of the family's income was spent on rent. All those who had girl children wanted to gather money for their daughter's dowry. Exceptionally, only one surrogate mother said she wanted to money to educate her three daughters. Educational capacity is known to enhance the surrogate mother's bargaining capacity within the surrogacy process (Pande 2010). All the surrogate mothers in my study could read and write, but none had studied beyond higher secondary level; two had completed up to the Twelfth Grade, six up to the Tenth, and five had completed primary education. Five surrogate mothers were domestic helpers, three were housewives, one was a clerk, one was a care worker, two were agricultural labourers, and another worked in the family agricultural land. Their family income ranged between Rs. 3000 and Rs. 6000 per month (Euros 50–100). The remuneration they received through surrogacy is equivalent to five years' worth of their salary. As one surrogate mother said, "It is not a fortune for us, but what we earn as monthly wages is spent on day-to-day expenses but here we get a lump sum" (SAMA and Sharma 2010).

The surrogate mothers did not receive a copy of their signed contract. This limited their capacity to file a lawsuit in case of breach of contract. They did not know the details of the Assisted Reproductive Technology Bill or their rights and duties under it, but they were aware that they were being exploited by not receiving a contract copy. Furthermore, given their socio-economic background, they might not even have the financial capacity to file a legal case. Second-time surrogate mothers have an advantage of a higher remuneration but this is based on the rapport they have maintained with the clinic. Accordingly, even if the intended parents were willing to solve all their financial problems, the clinic would discourage it because they are looking at the advantages of women repeating surrogacy and them earning more money out of the process. Hence, for the clinic, this is a profit-making practice and has little to do with eradicating poverty and helping the infertile people of the world. Second-time surrogacy has been an interesting phenomenon whereby, generally, men had left their jobs for a while and it was understood that the women would repeat surrogacy. It is only after the woman completed three surrogacies, that the husband would begin looking for work again. Exceptional to this is Nargisa; after her disappointing experience, she told her husband that she would be unable to repeat the surrogacy process and he agreed and returned to selling vegetables. They bought the house they were already living in and are content with their lives.

Rudrappa (2015) argues that the surrogate mothers are not from poor families as they earn Rs. 10,000 (<150 Euros). Being in this income quartile may indicate that they are not families living in extreme poverty, however, it does mean that they were on the edge of poverty and any unexpected event such as an illness, marriage, or death could slide them into abject poverty, especially in a country like India with no social security system. Hence, indeed most women participated in surrogacy to save their families from this. Nargisa's husband was a fruit vendor and could earn only Rs. 100 per day. The cost of living has gone up and this money seems to be just enough to meet the daily expenses. A nurse staying near their house saw their poverty and asked them if they would be interested in surrogacy. Nargisa earned Rs. 400,000 for the surrogacy and bought the house she had been living in, situated in a squatter settlement. She was confident that they could continue to earn a living with this income but she wanted some sort of stability. Buying a house could save them from paying rent which was Rs. 500–1000 per month. Sumita, also the wife of a vegetable vendor, did one surrogacy for buying a house and a second time to save some money for her children. One surrogate mother, Sarala was a care worker earning Rs. 900 per month and her husband's earnings were irregular. She was already involved in drug trials and other forms of biomarkets and thus was aware of surrogacy. She valued this as a quick money to uplift her family status. As her husband was resisting the idea, she involved her sister-in-law (Dimpy) too into this process to convince him that she would not be alone at the surrogate home. Dimpy and her husband were daily wage farmers and were convinced into surrogacy to buy a piece of agricultural land for themselves. Hence, the global demand/desire for surrogacy, the supply/need for money in India, the buying capacity of intended parents, profit making motives of medical practitioners along with the steadily increasing possibilities of technological developments are some of the important interlinks in this biomarket.

Selection Procedure

Courteous and submissive conduct was an important criterion in selecting surrogate mothers. Women who showed signs of assertive behaviour were politely rejected by the clinic on medical grounds. The medical practitioners expected potential surrogate mothers to be healthy, clean, understanding, and conscious of their responsibilities. The long waiting list of women willing to participate in surrogacy made interchangeability possible. The intended parents selected their surrogate mothers after face-to-face meetings at the clinic. For intended parents who did not speak the local language, the clinic provided nurses as translators who could speak both languages. The intended parents said they assessed their surrogate mother by her healthy appearance, willingness to relinquish the baby, family situation, husband's occupation, medical history, and family mortality history.

The intended parents also had religious preferences in selecting surrogate mothers. According to medical practitioners, 30% couples insisted that the surrogate mother should be of the same religion as themselves, which manifests their religious preference that can be interpreted as both/either personal bias and/or a potential discrimination. The Universal Declaration of Human Rights Article 2 states, “Everyone is entitled to all the rights and freedoms set forth in this Declaration, without distinction of any kind, such as race, colour, sex, language, religion, political or other opinion, national or social origin, property, birth or other status” (UN General Assembly 1948). After this selection procedure, the surrogate mother is asked to sign a contract, followed by the embryo transfer process. In case the surrogate mother did not conceive after one round of embryo transfer, she would have to go through the same procedure with another couple, since the clinic would not allow surrogate mothers to repeat the process with the same couple.

Role of the Medical Practitioners

The surrogate mothers in this clinic were expected to take care of the child after birth. Dr. Nisha trusted the surrogate mothers as the best caretakers of the babies, and they were paid a bonus for their services. Some intended parents wanted the surrogate mothers to breastfeed the babies, while others did not want them to breastfeed directly, so the surrogate mothers were asked to supply the milk using breast pumps. The surrogate mothers themselves were happy to do this for the babies. According to them, this was their only opportunity to interact with the children before they part with them forever. They said, “We cherished these days, as we all know, once these children are gone, they will never come back”. They took the babies’ pictures during this time and have preserved these. When I met them, after introducing themselves, the first thing they would do was to fetch the photos of the children they had borne through surrogacy. They sympathized with women who were unable or were restricted from this interaction with the babies. The surrogate mothers and the intended parents were ceremoniously called to the doctor’s office and in presence of Dr. Nisha, they religiously handed over the child (ren) to the intended parents. Surrogate mothers were expected to be rational individuals in control of their bodies and emotions at the time of relinquishment. One surrogate mother was scolded by a nurse: “Why are you crying now; didn’t you know this is what would happen when you started this process”. Caroline (an intended mother from Canada) said, “It is the clinic’s responsibility to prepare the surrogate and parents for the child’ s arrival and the separation, which is not done to the level that it should be. The clinics should counsel parents and surrogates and have to determine what kind of relationship each individual case is going to have. They can’t prescribe the same set of rules to each surrogate and parent and patient group. This reminds me of adoption in my country (Canada) and how it is either open or closed with respect to the relationship with the birth mother. Some want the openness and some don't; it's very individual. Clinic staff can help facilitate respectful relationships."

According to the doctor, bonding between the surrogate mother and the child is "Nothing but a false idea. The surrogate mother is prepared right from the beginning and taught that the child is not hers, and rightfully belongs to the intended parents. As a result, right from the beginning, the feeling of the surrogate mother towards the child is trivial". According to Caroline, "From the start, the doctors try to counsel the surrogate, making her aware that the baby had to be given away, as the embryo belongs to the biological parents (or donors)". Regarding psychological counseling the doctor said, "No feelings ever develop amongst surrogates for the child, so the question of resolving any feelings does not arise. The place of feelings for the child is taken over by the money in the case of the surrogate mother. In the entire process, she gives more importance to the monetary factor, while the feelings for the child are less. There might be some feelings towards the end, but as the memory blurs over a period of time, these too fade away. Finally, her feelings for the child become non-existent". Regarding the child's feelings, Dr. Nisha's opinion was: "Do you remember what or whom you saw when you [first] opened your eyes in this world? If you answer truthfully, it will of course be no. So no feeling ever develops in the child for the surrogate mother".

The medical practitioners controlled the payment scheme within the surrogacy process. The surrogate mothers were not paid until they had handed over the baby and completed all the requirements of the surrogacy process. The medical practitioners made sure that the baby was handed over to the intended parents only after the entire payment was made. All the payments made by the intended parents to the surrogate mothers had to be channelled through the clinic. The surrogate mothers were paid a monthly amount of Rs. 2,500 (Saravanan 2013). The first lump sum payment of Rs. 25,000 was made after completion of four months of pregnancy and again after eight months of pregnancy. The main payment of Rs. 177,500 was given to the surrogate mother after she handed over the baby. If there was a miscarriage at any stage or a stillbirth, surrogate mothers were not paid any extra amount. Dr. Nisha requested the intended parents an additional payment to be made to the surrogate mothers for breastfeeding and neonatal care services, or merely as a bonus. This sum was not fixed and depended on the intended parent's financial capacity. The surrogacy agent was paid Rs. 10,000 on successful relinquishment. According to Dr. Nisha, the surrogate mothers were generally not satisfied with the remuneration, "Eighty percent of the surrogate mothers are not happy with the money they receive. Compared to her work (as a surrogate mother), this sum is relatively less, but the economic background from which she comes is such that for her it appears to be the biggest gift in life". The couples spent approximately 20,000 euros (Rs. 1,200,000) for the surrogacy process, which according to Mary, an intended mother from Germany was doubled in case of twins. However, the payment for the surrogate mothers remained the same, except some money was paid as a bonus. The official amount paid to the surrogate mothers was Rs. 250,000 (Euro 3,430); however, this varied from case to case as they were paid extra charges for breastfeeding, tending to the baby and a bonus for additional children. The intended parents made this extra payment based on their capacity. In Ujwala's case, the highest payment made to one surrogate mother was Rs. 500,000 (Euro 7,143). Caroline wanted to pay more, but the doctor discouraged her, saying, "This could prompt such demands and unnecessarily raise expectations from other surrogate mothers as well".

Surrogate Mothers: Embodied Experience, Value and Reward of Reproductive Labour

Initially, on my enquiry to conduct a post-doctoral study from Germany, Dr. Nisha willingly gave me permission but when she met me face-to-face she became reluctant probably because she realised that I intended to stay there for a long period of time to obtain in-depth information about the respondent's lives. Unwillingly, she requested Dr. Harnish,[4] her assistant (also her husband) to help me. I would arrive early every morning but had to wait until most patients left, to be introduced to surrogate mothers. Most days I would just return to my quarters without any interview. At first, I was selectively introduced to surrogate mothers (Nargisa, Dimpy, and Sumita) and later met several other women through a snowball method and through the intended parents (Sarala, Ujwala Urmila, Maya, Sheetal). I say selectively because I noticed that I was introduced only to women who were seemingly submissive, soft-spoken, and were full of praises for Dr. Nisha, and looked up to her as a demi-god. With my extensive experience in conducting research on sensitive issues in India such as violence against women, working conditions in garment factories, and reproductive health, it was not difficult to read between the lines and observe their disappointment the between expressions of admiration. Any doubt that I had was cleared as I proceeded further with my research.

Nargisa

Eventually Dr. Harnish called me into his room after all the visitors had left and told me he would call a surrogate mother. After a while a young girl (19 years of age) slowly walked into the room along with a nurse. The nurse told her to answer whatever I ask her and she sat in the seat near me. She had the intravenous syringe still fixed to her hand. This was **Nargisa**, she had delivered twins through caesarean, a girl and a boy for a NRI (Non-resident Indian) couple just two days ago. "With twins, there was no chance of a normal delivery," she told me. But I was to

[4]Dr. Harnish was an orthopaedic doctor himself but Dr. Nisha's business was thriving because Aasha Clinic had become a globally popular surrogacy destination. So he decided to assist her instead of practicing his own profession.

know later that each and every surrogacy birth in this clinic was a caesarean. During the conversation, she also told me that the intended parents were from Tamil Nadu but had settled in Australia to which I informed her that Tamil is my mother tongue too. I didn't want to take much of her time as she had just delivered. I asked the whereabouts of the children, she told me they were with the intended parents at a hotel. I then enquired why she has not returned home. She said, "It depends on the passport, visa clearance. I can go home only after they (the intended parents) go because I am still feeding the babies". She continued, "As the children may get infection, they prefer to keep them with themselves in a hotel nearby, the husband comes two times a day to fetch my milk." I then asked if she wants to keep any relationship with the intended parents to which she told, "They have told me so, I have given them my address and phone number and I have told them don't send me anything but just send me the photos of the children every year, I want to see them as they grow."

Within the clinic, the top floor had a few dormitory rooms and some private rooms. The surrogate mothers were not allowed to use the stairs. They could move between floors only using the lift and with the assistance of the nurse who operated the lift with a code number, "because it is other people's children we have to be extra careful". Her family consisted of her husband and their four-year-old son. Nargisa's husband is a fruit vendor and could earn only Rs. 100 per day, but as she explained, the cost of living is increasing and this money is barely enough for daily expense. She was confident that they could continue to earn a living with this income but she wanted some sort of stability. Buying a house could save them from paying rent which was Rs. 500–1000 per month. The amount she was told for the surrogacy was 3 lakh, but she said "Madam asks for more after birth keeping in mind our poverty". She also added, "It's a blessed deed '*punya ka kaam*' because I get some money to buy a house and someone gets a baby they want". Some scholars have written extensively about surrogate mothers considering surrogacy as "work" (Pande 2010, Rudrappa). However, I would like to bring to the notice that in vernacular Indian languages, the word *kaam* is also used to express endeavour, deed, or an act. By saying "*punya aka kaam*" the surrogate mothers actually means "pious deed" or good karma.

When she first came to the clinic, she was explained about the surrogacy procedure and both she and her husband were required to undergo a blood and urine test. Nargisa had to stay at the surrogate home for several months. She resided very close to the clinic, so she requested the doctor to allow her to go home on the pretext that her in-laws were creating problems due to her absence, and she was allowed to go home. She persuaded the doctor that her community was conservative and would stigmatize her if she was away from home for a long time; hence, she was allowed to go home between the second month and fourth month. When her pregnancy started showing she came back to the hospital to avoid social stigma against surrogacy. She says she was one of the rare persons who were allowed to go home, because of her quiet and submissive nature. Every Sunday her husband and son were allowed to meet her. She was restricted to the first floor during her entire pregnancy and came down only for check-ups. During this meeting, we had a very basic conversation about what motivated her to be a surrogate mother, from where did she hear about this, and how her surrogacy experience was. But I met her again coincidentally in the surrogate home which turned our mere acquaintance into a long-term relationship; I still have contact with her and call her once in a while. At this first meeting she told all positive things about the experience and was hopeful that the couple would keep in touch with her.

I came to the clinic 10 days later to meet Dimpy (see case study below), and I coincidentally met Nargisa. She was standing in the corridor, the syringe still on her hand and seemed distressed. I stopped to speak to her and she asked me, "Can you please do me a favour? As you know Tamil, can you talk to the intended father when he comes later today to collect my milk. Can you please request him to allow me to go home? I don't want any extra money, I just want to go back home, I am fed up of staying here especially when the children are not here with me. At least I can go back to my son. I can't tell the doctor, I am scared. Can you please talk to the (intended) father directly, maybe he will listen to you?" She was also apprehensive and uncomfortable in using the breast pump. She asked me if there were any side effects in using this machine. I waited for a few hours and when the intended father showed up I took him to the corridor and told him about Nargisa's request to which he replied, "Let me see, I will try my best to let her go". But Nargisa was frustrated when they promised to let her go home but continued to keep her there for four more weeks on some medical pretext. Although I had little hope from this person, I requested him (the intended father) if he would be willing to share his experience with me, to which he replied, "No, its controversial, and it's illegal in Australia so I don't want to get into trouble, you know what I mean."

Nargisa was paid Rs. 400,000 finally, the surrogate mothers are told this is a bonus amount but it is actually the money paid for breastfeeding and a part of the double payment made by the intended parents for twins. I went to her house a few weeks later. She came to pick me up at the station and told me to wear a headscarf and pretend that I was her husband's friend because she was not supposed to have a friend who clearly seemed to be not a local. Despite this overall conservative environ, I saw a very different Nargisa, she was not the soft-spoken young girl I met at the clinic, she was the eldest daughter-in-law in the family and she played a mother role to her younger married sister-in-laws. That day there was a quarrel between her sister-in-law and her husband and she played an important role in solving the conflict.

Another day when I was at her house, she told me, "You know Sheela Ben, my hair has started falling in the last few weeks." She paused and then continued, "I think the children have started teething, we believe that when our babies start teething, the mother loses her hair". Last year I met her again and asked her if she would repeat surrogacy to which she answered, "Never again, *usme bahut magajmaari hain* (it's too complicated)". I had heard the doctor speaking in a TV interview about how she helps the surrogate mothers with school books using a surrogate mother's trust fund she had set up. When I asked Nargisa she said, "I never go there, she's (the doctor's) giving away cheap things that we can easily buy from the shops ourselves, why should we go there for taking those books. Women who are extremely poor may go. Maybe if she pays for school uniforms it would be useful".

I met her several times thereafter and have had telephone conversations with her for the last seven years. She bought the house she had already been living in, situated in a squatter settlement. When I interviewed Nargisa inside the clinic, she described the surrogate home as "like my own house" but it was only when I met her outside the clinic environment that she openly told me how selfish she felt the clinic and the intended parents were in the rules that were imposed on their movement within the surrogate home, the food they ate, and, in her case, not heeding to her pleas regarding breastfeeding and requests to go home. Regarding the intended parents she said, "It has been seven years but they have not called me once; *Ehsaan Faramosh* (ungrateful)". Since these seven years, there have been various offers to exchange her phone; she could have changed her phone and sim card, but she has kept her old phone and the number with the hope that they will call one day and let her know about the well-being of the children. I am one of the few people who still calls her on this number. Whenever I call her from a new number not registered on her phone I can sense the anxiety in her voice as she has maintained this phone only for the long-anticipated call from the intended parents.

The case of Nargisa brings forth the cruelty and inhuman way that intended parents and medical practitioners deal with surrogate mothers. Her emotional need was secondary to the desires of the intended parents; she was forced into using a pump to extract her breast milk, although she was finding this extremely uncomfortable and despite showing a lack of interest on the extra money given for this purpose. Her urgent need to return to her son was ignored. The intended parents cheated her by assuring to keep in contact with her. Her embodied connection with the children was revealed in her relation to the changes in her body to the children. Her lack of agency within the clinic environs is revealed in her inability to express her requirements to the doctor and her inability to speak directly to the intended parents due to language barriers. This lack of agency is because of the surrogate mother's vulnerable situation as well as the hierarchy.

Dimpy

After my first meeting with Nargisa, I continued my routine waiting at the clinic. Dr. Harnish called me inside his room one day and asked me if I would like to meet a surrogate mother who had been shifted to a children's hospital along with her new born baby two weeks ago, and was waiting for the intended parents to arrive from Turkey. There was tension in the clinic regarding these Turkish intended parents as they were unable to contact them. The doctor's assistant told me, "We cannot trust anyone other than the surrogate mother for the baby's well-being." He gave me the directions to the hospital and the room number. On reaching the children's hospital, I realised it was a small clinic with no reception. There was a ward-boy on the corridor and I told him that I had come from Asha Clinic and have been sent here to meet a Dimpy. He directed me to a room and as I entered I saw Dimpy, her husband, and another caretaker (a nanny), and a little baby girl in the room. To my astonishment, Dimpy had not even been informed about me or about the purpose of my visit. I introduced myself, explained the purpose of my visit, left the consent form with them in order to give them time to consider if they would like to participate in my study. I realised that this was a pattern not only in this clinic but the other clinic in Ahmedabad too. That consent was never taken from the surrogate mothers, it was rather an order to answer whatever I ask along with a set of instructions about what they should and should not discuss with me. I soon received a call from them saying they would be pleased to participate in the study and requested me to visit them the next day. In the following days, I became a regular visitor to the hospital. Dimpy was fed up eating the same lunch box every day and I would buy something different for them to eat. She and her husband (Dhiraj) had been looking after the baby girl since she was born (for 2 weeks), breastfeeding, changing nappies, and providing all other required care for the child. Dimpy's attachment to the baby was very evident in her affectionate kissing and fondling, either in response to distressed cries or during play time. She told she had started breastfeeding just naturally as the baby was with her all the while. The couple had even named the child as Amita (also their younger daughter's name). I also became very fond of little Amita as I spent more and more time with them. They were very hopeful that the couple would want to keep further contact with them.

Dimpy and Dhiraj were agricultural labourers from a neighbouring rural area. She was convinced into surrogacy by her husband and sister-in-law and they agreed because they wanted to buy a piece of agricultural land for themselves. Dhiraj came to know about surrogacy through his sister Sarala. She was a nurse and wanted to earn extra money through surrogacy as she was staying in a quarter provided by the church. She persuaded her brother Dhiraj who then persuaded Dimpy to go with her for the surrogacy as Sarala's husband was unwilling to send her alone to the surrogate home. Although both of them began the surrogacy process together, Dimpy was unsuccessful in the first attempt, while Sarala became pregnant at the first attempt. Dimpy's first intended parents, she says, were very nice people, they were Afro-Americans. They paid her well and spoke to her kindly when they met her for the first time. They also gave gifts to her children. The second attempt was successful with a Turkish couple. This clinic does not allow a second attempt on the surrogate mother using the same set of gametes. On a positive pregnancy, Dimpy joined Sarala, who was already in the Nadiad surrogate home.

Dimpy and Dhiraj slowly began sharing with me about their children and how their boy excels in studies, the problems they are presently facing by leaving them at home with relatives, how they must depend on them, and why they have not informed their family about the surrogacy. They have kept this secret not only because of the probing questions but also because they do not want to alert them about the sudden cash influx and attract unnecessary scrutiny into their lives. As it is a rural area they were also concerned about how people would react. She and Dhiraj spoke to me freely as this place was not a surrogate home nor was it within the clinic environs, it was a children's hospital a few kilometres away from the clinic. She told me about some of the violating practices within the contract and within the clinic. The surrogate mothers are not given a photocopy of the signed contract lest they take any legal action against the clinic, "anyways even if we get a copy of the contract what difference does it make, we don't know anything about our rights by law". She told me, the selective abortions in utero are conducted on behest of the intended parents resulting sometimes in abortions of all fetuses. She couldn't understand why they would implant five embryos into a surrogate mother if they want to limit the babies to only two. Although legally only three embryos are allowed to be implanted into the contract mother's womb in India, up to five embryos are known to be implanted with a high likelihood of multiple pregnancies. In case of these multiple pregnancies (triplets), the doctor suggests "selective reduction" of one or more fetuses. This procedure may also result in miscarriage of the remaining fetuses, preterm labour or infection. The preference of the contract mother is not asked either for the number of embryos to be implanted or in the decision making about selective abortions. Although the success rate of the gestational surrogacy is considerably low, the contract mothers are not given any compensation if they experience a miscarriage at any stage of the pregnancy. Studies reveal that the overall pregnancy rate per cycle after IVF surrogacy was only 24%, with a clinical pregnancy rate of 19%, and a live birth rate of 15.8%, while the clinics usually publish an exaggerated success rates on their websites (Goldfarb et al. 2000). In the surrogacy contract, women have to sign off all rights on medical interventions.

Dimpy had lied to her family that she was working in a factory that wouldn't allow her to return home. However, they grew suspicious when she never returned even for her mother's death. She also told me, a normal birth is not allowed with surrogacy, even if the surrogate mothers get pain, they are rushed into the operation theatre for a cesarean. Dimpy didn't know the intended parents very well, she had only spoken to them over the phone with the assistance of an interpreter from the clinic. They had met the intended parents only once. The couple arrived from Turkey when baby Amita was 21 days old, they had brought her some gifts and after thanking her profusely they took the baby along with them. I was not present that day as I had to go to Ahmedabad. When I returned, Dimpy had been shifted from the children's hospital to the surrogate home above the clinic. She had to wait here until she received the money. Dhiraj said, "We cannot keep coming here to collect the money, we live far away". A few days later, I was sitting alone with them when a nurse came to the room with a bag full of cash. The couple had come, paid the money to the doctor and had gone without meeting or saying goodbye to Dimpy. Dhiraj closed the door and opened the bag excitedly and told me "I have not seen so much money at a time in my entire life". The entire transaction was a cash dealing. He started counting the money, he gave me a few bundles to count, the total amount was three hundred thousand rupees. It was more than what was written in the contract. Dimpy was evidently disinterested in the money or the counting, but she was happy that her husband was happy. Soon they began packing their bags to leave for their village; she was going to meet her children (4 and 5 years of age) after a year. She was happy albeit at the same time worried about what her in-laws would say to her. They invited me to their house and asked me to reach their village a few days later so that it doesn't arouse any suspicion about Dimpy's whereabouts for 12 months. It was Sarala (Dimpy's sister-in-law) who took me to Dimpy's village and her son came along too.

There was absolutely no psychological support given to her. Although she appeared to be calm, but knowing a little about her by then, I knew that she was going through emotional turmoil. She told me, it's better to be calm as she had seen another surrogate mother being scolded by a nurse[5] so she preferred to keep her feelings to herself. She had spent the last year only experiencing this baby growing inside her, she was also given the responsibility of taking care of this baby for 21 days, and now she felt lost. She was feeling detached from her two children back in the village with whom she had no contact for the last one year, and with baby Amita, with whom she had bonded and had spent the entire year expecting had abruptly been taken away from her. As the intended parents walked away without even meeting them for one last time, their hopes of keeping in touch with them and baby Amita was shattered. Dhiraj told, "We are going to erase her (Baby Amita) memory from our lives, we will never mention her name from now onwards". This was almost a warning for Dimpy and for me too, never to speak or ask about Baby Amita. At this point I ceased to be a participant observant but turned into a participant in the process myself because his warning was directed to both of us (me and Dimpy) and not just towards Dimpy. Although this was supposed to be a question on my research aims, I was not allowed to mention Amita's name ever again.

A few weeks later, when I was at their farm, taking photos of him, Dimpy, and their children, and as usual after taking the photos we (myself and Dimpy) were scrolling through the photographs on my digital camera, I over-scrolled accidentally exposing one of the recently taken photo of Baby Amita and both of us froze; firstly, because nobody among her in-laws knew about Baby Amita, and secondly, because her husband had warned us to erase her from our memory, never to mention her name again (Photo 5.1). We looked at each other, I could sense the tears swelling up in her eyes, she paused for a few seconds, and heaved a deep sigh before turning and getting on with her mundane household work. I had by mistake hit a sensitive note; she was supposed to be alive only in our memory and never to be expressed. That's when I realized that I too had bonded with Baby Amita as I met her every other day, played with her and helped them care for her (Photo 5.2). I accompanied Dimpy to her children's school where she had to face the blame from the teacher who scolded her for not being around for her children for the past one year. They asked her where she had been for one year, that there were rumors that she had been to a nearby town, and was questioned why she couldn't come to see her children. Dimpy had been an egg donor before and she returned to this

[5]Moreover, in general nurses in Indian hospitals, especially in maternity clinics, asking women to "cut the drama" in case a pregnant women panics, is a very common phenomenon. So surrogate mothers are aware of this and research on institutional access to delivery reveals that women want to avoid bad treatment in hospitals and hence prefer to deliver at home.

Photo 5.1 The photograph, the author mistakenly scrolled into

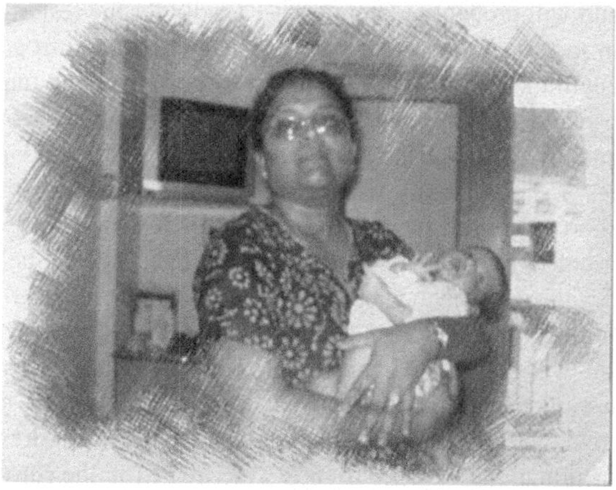

Photo 5.2 Author with baby Amita

practice to earn some money. She had shared her experience of severe health problems she had faced due to egg donation.

Conclusion: This case brings out the importance of the mother–child bonding during the gestational period and thereafter. In this case, I too bonded to some extent with the baby as I was visiting the children's hospital every day to spend time with Dimpy. The baby had started responding to me as a familiar face. Some scholars have found that surrogate mothers do not bond with the babies. The medical practitioners in the clinic say that maternal–infant bonding is an illusion and it doesn't exist. In this case as I had also embodied a bonding with the child, I can say with conviction that this bond is not an illusion. According to Dr. Nisha, we keep reminding them (the surrogate mothers) that the baby looks like the couple, and say, "This is the couple's baby, and when it's a foreigner, she (the surrogate mother) cannot relate to the baby as it is Caucasian white and she is Indian. She cannot relate because we do not use the egg of the surrogate (mother)". Such a comment about a perceived alienation of a surrogate mother's bonding with the child based on the skin colour rather than stating the reality was geneticised racism. Why ever would a human with brown skin not bond or not want to bond with a child of another skin colour? She added "In my experience 99% have never had had any problems, 1% may remember for a few weeks". Needless to say, not only Dimpy, even her husband and I who had for some time cared for the child had bonded with her (refer picture of me holding Baby Amita). Her skin colour had nothing to do with how Dimpy related to her.

Dimpy and Sarala were already involved in egg donation and it was through this that they were informed about surrogacy, and her motivation for surrogacy was to earn extra money to buy some agricultural land. She had been repeatedly told by Dr. Nisha and others in the clinic throughout the pregnancy that she was just renting her womb by carrying someone else's baby to term only to return it at the end of the process in return for money and yet, she happened to bond with the baby, name her, and will probably remember her fondly for the rest of her life. The intended parents in turn have been told by Dr. Nisha that the surrogate mothers don't want to do anything with the child so that they don't feel any morose when they take the baby back to their homes. This hence brings out the complexity of this relationship and that relinquishment is not as straightforward as reported by many liberal researchers. Dimpy was rewarded for her reproductive task in bearing and, for a short period, in rearing the child. She felt valued by her husband as she had contributed towards the betterment of their lives. She and her body were grossly devalued by the medical practitioners and the intended parents who used her body to fulfil their financial and social motives. She expressed feelings of being used when she was not given a copy of the contract, although both she and her husband could not understand what was written in it. She was not happy about living in the surrogate home for nine months away from her children. She expressed her dissatisfaction at the way her friends were treated in the surrogate home in being forced to undergo in utero selective abortions and compulsory cesareans but she couldn't express this to Dr. Nisha. Selective abortions in utero could lead to abortion of all babies. If she expressed her concern, she says, they would call her to the clinic for selective abortion and purposely do an abortion on her. She says they live in this fear as the clinic has complete control over their bodies and the surrogacy.

It was her substandard socio-economic status as a woman that pulled her into this practice, persuaded by her husband and his sister, Sarala. Whether she was a reluctant contender and whether the persuasion of her husband and sister-in-law was stronger than her opinion is unknown. The husband did play a strong role in determining the erasure of the little girl from their life which was in contradiction to what she would have desired or what she had experienced with the baby. It was not a primary concern for Dr. Nisha nor was it of any consideration to the intended parents. Dr. Nisha did not interfere with how they would spend the money but she advised the husband not to bring his wife again for surrogacy knowing very well that this first time was an exploitation of her body. They had not done this on account of abject poverty and they were managing well even without this extra money. Dimpy expressed her discontent towards the clinic and Dr. Nisha's money-minded intentions but she didn't share the same discontent about her husband, because I was never alone with her, he was always present along with her everywhere.

Sarala

It was Sarala (Dhiraj's sister) who took me to Dimpy's village. Sarala worked as a nurse 12 h a day and earned Rs. 900 per month. Her husband worked as a gardener in a Christian Missionary and his salary was unstable. Some months they paid him all the arrears together and other times he was not paid. She came to know of surrogacy through some friends. She was unable to convince her husband that this would be a good option for her. She then persuaded her brother Dhiraj and through him his wife Dimpy to go along with her to the surrogate home. Both women from the household set out to earn money through surrogacy. Both went to the clinic together and went through the initial clearance and met the respective intended parents. However, Sarala became pregnant at the first attempt to a couple and remained at the surrogate home, while Dimpy was unsuccessful and had to return home only to join her again a few months later. Sarala's intended parents were non-resident Indians (NRIs), originally from Rajasthan. She received Rs. 350,000 for the surrogacy. Sarala has three children, two daughters and one son. She had left her two children in a Christian missionary hostel for nine months because her husband was unable to look after them and her mother-in-law refused to care for them. She was worried about their well-being but couldn't go to meet them although she was staying in a dormitory in the same town.

She described her surrogacy experience as: "This process is so distressing that I would not have done it even if someone paid me 10 times the remuneration, had I been well-off, but I am so desperate (for money) that I would do it even if I was paid just one third the amount". But after her surrogacy she became a surrogate agent persuading other women into becoming surrogate mothers at the clinic for which she received Rs. 10,000 which would be paid to her directly by the surrogate mother after she had received her payment. On a causal trip to a surrogate home along with her to meet another surrogate mother she had recommended, this surrogate mother scolded her, "You never told me that these injections are so painful, now I feel like running back home", to which Sarala replied, "Are you not getting all that money in return"?

She (Sarala) had to stay in a dormitory home at Nadiad close to her house and came home only for festivals as she had good relationship with the matron. Her sister-in-law joined her at the clinic after a few months. Her brother and her husband would visit them every weekend and brought them some home food. Sarala resisted intermediary power such as the dormitory matron and tried her best within the home to improve the food and facilities. Sarala said the only time they cleaned the dormitory well was when any documentary crew came with cameras. Intended parents never came to this dormitory as it was situated around 20 km away from the clinic and only a few documentary crews came here. Sarala said the food provided at this dormitory was sub-standard although the intended parents paid Rs. 6000 per month per surrogate mother for this purpose and there were at least 10 surrogate mothers housed at this home. On enquiring about the food, the matron would say, "At least here you are eating two full meals a day and not just the *Baakhri* and green chilies that you eat at home so be happy with whatever you get". Sarala complained about the food to Dr. Nisha but it was to no avail. Sarala, however, told me never to disclose what she told me to Dr. Nisha as she still works for the clinic as a surrogate agent. Sarala's husband complained that the intended parents are not bothered and never try to find out about what happens in the dormitory homes and they should take more care of what the surrogate mother eats and in what condition she lives. One intended parent (Caroline) told me that it was not her responsibility, but the clinic's to look after the surrogate mothers (see Chap. 6). Some of the cash and other gifts that were sent by the parents were also generally pocketed by the matron. Sarala said phone calls from intended parents were not given to the surrogate mothers in order to keep a control over their relationship. She gave birth to twins, one boy and one girl and she showed me their photographs, both were grossly underweight (Photo 5.3).

Overall, Sarala had a bad experience with the intended parents after birth. She tended to the babies after birth as the parents arrived late and she was also breastfeeding the child. After the parents came and took the babies to their hotel, Sarala too left for home without waiting for payment. She was also eager to see her children and husband who had been without her for a year. After a week, she called up the parents and requested if she could see the babies and they agreed. They asked her to wait at the clinic in the evening where they would bring the babies. She waited there patiently expecting them to come. She even called them twice and she was assured that they would bring the babies but they never came. She waited with her son and returned home very late at night, feeling dejected. As in many other cases, the parents never called her to enquire about her or ensure the well-being of the children. The behaviour of the intended parents changed from kindness (during pregnancy) to neglect (after relinquishment) and this was observed by all surrogate mothers. Sarala remarked, "In a way it is good that they are not in touch, otherwise I would have asked for the children to be returned".

Photo 5.3 Child born through surrogacy: grossly underweight (Photo taken by author on 17 October 2009)

Sarala is a complex person who felt exploited, violated and degraded as a surrogate mother and yet coerced other women into surrogacy without telling them the whole truth about her experience as a surrogate mother. When I told Sarala about Gargi who gave birth to a child in Ahmedabad, she came along with Arpita to visit Gargi at the hospital to offer solidarity and appreciation for having the courage to be a surrogate mother and then part with the children without seeing their face.

In general, surrogate agents combed the poor-income residential areas in search of women facing financial hardships or family problems and convince them to become contract mothers. They also searched for desperately poor women in drug trial centres in the lookout for women and families who are already involved selling their body. These families tend to accept the proposition of surrogacy more easily than those who are unaware of biomarkets. One of the most important pieces of information conveyed to surrogate mothers was the clarity about chastity in the surrogacy process. On being thus informed, women convinced their husbands. Women generally also try to convince at least one friend or a relative to go along with them into the surrogate homes for the first time as they are scared of the unknown (Saravanan 2013). Where men first received the information, they have convinced their wives and other women in the family into this practice. In my field research, I didn't come across any women who had to convince their husbands for repeating the surrogacy process. Some men even quit their jobs and coerced women into repeated attempts of surrogacy. There are no studies that have been conducted as yet to know what becomes of women's agency after their reproductive capacity becomes redundant.

Sumita

Back at the clinic, Dr. Harnish again came out of his room and asked me if I would like to join some others who were on their way to the surrogate home in the town to attend a baby shower ceremony. The baby shower was for a Japanese man and his surrogate mother, the intended mother was nowhere to be seen. I had only visited the surrogate home within the clinic but had never seen the dormitory homes, one within the city and one several kilometres away. I met several surrogate mothers at the surrogate home and also a few intended parents, but only Caroline accepted to participate in the study and she promised to give me a few contacts of German parents as she knew I had come from Germany. From then onwards began my semi-structured interviews and participant observation with the intended parents (Caroline, Shyamala, John, Mary and Jignesh, see next chapter).

At the surrogate home, preparations were being made for a religious ceremony on the ground floor; some surrogate mothers were upstairs. They were not allowed to attend this ceremony as it was their first three months after implantation during which they were required to take complete bed rest. In the veranda were two intended parents who were sitting and chatting and there was a documentary film crew who were supposed to arrive soon. Within the surrogate homes, beds were lined up in a hostel-like environment. In my following visits to the surrogate home, I was told there were restrictions on their movement not only immediately after embryo transfer but also throughout the pregnancy. They were not allowed to use the staircase.

Sitting in the waiting area of the clinic was in itself an ethnographic experience. It was a very busy IVF clinic in a small town in Western India. Intended parents from abroad would arrive in air-conditioned cars and were taken to another secluded waiting room also with air-conditioning, with clean toilets. Everyone had to leave their footwear at the entrance, but some intended parents carried their footwear along with them, fearing it would be lost. They were soon called into the doctor's room and within half an hour they returned in the same car to their hotels.

The surrogacy experience of Gargi and Ujwala are included along with intended parents. As I mentioned earlier, this was a participant observation method in which I almost became a part of their family, hence, a very time-consuming process. One such one-off semi-structured interviews I conducted at the clinic was with Sumita. I was introduced to her in one waiting routine to meet surrogate mothers; she was sitting in a room with her husband (Raysingh) and son (Tushar). This was Sumita's second surrogacy; the first time it was for a Japanese couple and this time it was for a non-Hindi speaking Indian, so she couldn't converse with them. "This time they are Indians, but they don't speak any Hindi, they speak only English. If they speak Hindi, I can understand but if they speak English I don't understand. Most of us in Gujarat can speak Hindi." I asked her if she liked the couple, and her reply was "*Haan, pasand kiya mujhe*", (yes, they liked me) sounding like a bride-choosing custom in India when eligible bachelors go house-to-house selecting girls for marriage on the most easily visible attribute, after the initial selection has been done on the basis of education and family background. I asked her again, "did you like them?", to which she answered, "Yes, I liked them, they gave chocolate to Tushar." She could not bring herself to imagine herself being a person with enough agency to be able to select intended parents. She continued with why she was selected, "In my previous surrogacy, the baby was four and a half kilos, so these people selected me because I am fat (*moti*)," hoping that she would again delivery a bigger baby. She hopes a bonus amount for being fat, "Moreover, I am a *Patel (a landowning caste)* so I will be paid more than what they pay others". Another surrogate mother, Arpita was rejected because she was too skinny and the intended parents were concerned if she could bear the procedure. Dr. Nisha explained to them that housemaids are usually very thin and moreover she told them that this procedure is meant for women like Arpita who were in dire need for money and they should agree and eventually they agreed.

Like all other surrogate mothers, Sumita too complained about how she had delivered all her children at home through normal delivery: "None of my children were born in the hospital. I was working until 8 and around 9 p.m. I had my first child, with my second child I had pains at 12.15 and in 30 min the child was out. But here they say they had to do caesarean because the child was big". There is an increasing trend for caesarean-section (c-section) in India as a medical intervention and a conscious decision taken by upper middle-class Indian women. In this case, women are forced into a c-section by the practitioners. Just like everything else in the surrogacy process is managed, birth too is medically managed. According to Julie Bindel's research in Anand, Gujarat, one intended parent from the UK who have five children through surrogacy had insisted upon a caesarian birth with all of his children's births because they didn't want any of their children coming in contact with a woman's vagina because they found that so disgusting.

Sumita's family earnings was Rs. 3000 per month, they earned Rs. 100 every day selling vegetables on a trolley, they would buy the material for Rs. 300 and sell it for Rs. 400. According to Sumita and Raysingh, they need approximately 10 lakh to come out of poverty, they earned 2 lakh from the previous surrogacy and expect 3 lakh from this surrogacy and some additional bonus. This, according to Sumita meant she would have to go through at least one more surrogacy and they would be closer to their target and Madam (Dr. Nisha) will help them with this. They bought a house with the first surrogacy and with the second one she wanted to save some money. She wants her son to become an engineer, "We spoilt our lives, at least we will educate our child." Nowadays by giving one lakh rupees, one can get an office job. For the girls, apart from the dowry people have to give a sofa set, fridge, and so on. She says, "That is why we have to do this (surrogacy)". Her husband said, "When I was in the third grade my parents died and I couldn't continue my studies, but all my three brothers and one sister became rich, we are the only ones who are very poor." Her mother-in-law is impressed with her since she became willing to become a surrogate mother. Sumita said, "My mother-in-law says, my poor son, how can he ever earn that much money in one year with his present work, so my daughter-in-law is doing good, she is doing it for my son, that's how my mother-in-law thinks". Regarding stigma, she says, "Everybody talks behind my back, let them talk I don't care. I have to think for myself, not about anybody else. My parents say, they have given me off in marriage, now whatever I do in my life is between me, my husband, and his family members, they have no say in it. My in-laws invite me to their house so I am happy, if they had excluded me I would be very disappointed".

I asked how the little boy (their son) stayed without her when she went into the surrogate home for the first time. "He was one year old, I was breastfeeding him at that time when I moved into the surrogate home." However, she requested the intended parents (a Japanese couple) if she could go back home for some time because of her small child. "I convinced them (the intended parents) that I will not let anything happen to your child. I will take any mishap onto myself and would let something happen to me rather than to your child and then they let me go to my little baby at home." This follows the unwritten stipulated rules by the clinics in India, that in case of life threatening situations, the baby will be given more priority than the mother. This prioritizing of rights maintains the hierarchy in the value of life between the rich–poor, global North–South.

In the first surrogacy, she gave birth to a baby girl, she breastfed the child for five days after which the intended parents decided to move into a hotel with the baby but she continued supplying milk using a breast pump and after 15 days they left. They called her once in the last two-and-a-half years and sent her photos when the baby was one year old. I asked her what she would do this time as her child was still small, "We have to do whatever they want us to do, if they ask me to stay at the surrogate home, I will stay here, if they allow me to go to my house I will go back". At this point, the husband intervenes, "It's better if she stays here, I will come every week with my son, show him his mother's face from the entrance of the room and convince him by saying see your mother is sleeping there on the bed, she is sick like all other women sleeping in the room and if you go to her you'll fall sick." With a sign language of embracing, she told me that the surrogate mothers were not allowed to hug children, they could just see them from a distance. The husband added, "I told my child the mother has fever, look there, she is on the bed, if you go near her you will get the fever too." I asked her how did she manage to stay away from her little child for several months, to which she said, "If I think about attachment, I cannot focus on my goal. If I allow him to cuddle me then he will say, Mummy I want to stay with you, I want to eat with you, I want to sleep with you, which is not allowed here, that's why I keep him away." He (her son who is now four years) was intently listening to our conversation sitting in the room. Surrogacy for her was a duty, a goal towards accomplishing her family requirements, her husband also spoke about her being at the surrogate home as an irrefutable condition that she and the child should adjust to, although he was too young to understand. They had referred only to their son's education and their dream to make him an engineer, while for their daughter they aimed at sending her to school up to the Twelfth grade so that even if she gets a minor (*chota-mota*) job, she can be independent. I couldn't help myself and advised them to educate both the son and the daughter equally. At that point, Dr. Nisha entered the room to use the washroom although she has a washroom in her room too. As soon as the doctor enters the room, the husband showers praises on her: "Madam is very loving to our children and to us." This pattern of praising the doctor in the middle of the interview in case someone entered the room or passed by was something I had become accustomed to by then.

Uma

Uma left her drunkard husband because he was violent towards her and her son; he would beat her to grab the little money she earned working as a housemaid to spend it on alcohol. She left his house to live with her brother and sister-in-law. In their house too she was considered a burden and faced abuses. She got to know about surrogacy from one of her friend also working as a housemaid and she took her to the clinic. She explained the surrogacy process, that the genetics of the intended parents will be used to create fertilised eggs in the laboratory which will be placed in her womb and she will have to carry the baby to term for which she will be paid Rs. 2000 per month. On giving up the child(ren) born she will be paid the remaining amount Rs. 212,000 (after deducting 18,000 of monthly payment and 10,000 of surrogate agent's fees). She would be given the first installment of Rs. 25,000 after the fourth month and another Rs. 25,000 after the eighth month. When she met the intended parents, they asked her, "Would you take care of our child(ren)", to which she answered "I will take even more care than I did in my own pregnancy". Regarding the surrogate home, she said "I was told about the conditions about staying in the dormitory house and that I will not be allowed to go home; I was unhappy about that but signed wherever they needed my signature". When the process began, she said, "The injections were so painful that I wanted to quit and go back home, but thinking about the future of my child I felt I will have to tolerate this pain". The dormitory house at Nadiad was in a bad condition, no water to take bath, the food was bad, and this was partly because the matron pocketed 50% of the money she was given to maintain the house. On complaining about the situation, the matron would treat the particular surrogate mother badly. Uma said, "This nine months of '*takleef*' (bother), is nothing compared to the money I will get in return and the future of my child." During the time in the dormitory house she was always worried and would cry thinking about her child as she had separated from her husband. Her six-year-old son was staying with her brother and sister-in-law; she said, "If my child needs anything, how will he ask my brother? Children can ask whatever they want without any hesitation only to their parents, moreover he is small. How will he get ready to go to school every day?" She was moved into the dormitory home above the clinic at the ninth month and a caesarean happened soon thereafter.

She experienced post-partum haemorrhage and despite several attempts it couldn't be controlled and so she was operated upon again for the removal of her uterus. The child she gave birth to was also unwell and was admitted to a pediatric hospital, she didn't know the exact health problem of the child. Uma was shifted to the pediatric hospital soon after she felt better to be able to breastfeed the child. The intended parents were from the UK and they had no interest in keeping in contact with her. The questions that came to her mind were: "I have no right whatsoever over the child whom I carried for 9 months, I have to give away this child with complete knowledge that I will never be able to see him ever again and this child will never know that the surrogate mother who gave birth to him, almost gave her life in the process, my child was born through my womb and this child was also born through me but they will never meet each other and never know that they are half-brothers". She says, "It is not just me but all mothers who go through this dilemma when they have to part away with the child." The day she had to hand over the child, she was called into the doctor's room and the intended parents came there too. She says, "I cried and cried and I felt as though I was handing over a part of my body to them, I felt as my soul is parting from my body. But it was my duty as a surrogate mother that I had to abide by, if I had to get the final payment. The intended parents left with the child and I too left for my brother's house after taking the final payment". The questions that came to my mind were, if Uma was protected in her informal employment as a housemaid, if she had easy access to a loan of Rs. 240,000 (USD 3,700), she would not have had to experience a near-death situation and would not have to experience the psychological trauma of parting with a child that she considered as much hers as the child she had left behind at home; she would probably live with this pain for the rest of her life.

Shama

I was introduced to Shama in the second IVF clinic without a surrogate home. She was five months pregnant and also a surrogate agent, and had brought several women as surrogate mothers into the clinic. I was also introduced to Gargi in this clinic to know that there were cameras in every room in the clinic so that the clinic personnel could keep a watch over the surrogate mothers. I conducted both these interviews within the clinic, the difference being Shama did not share her contact with me as advised by the doctor but Gargi found a way out to be able to share her contact with me (see Chap. 6). Her intended parents were from India but could not speak Hindi so she could not communicate with them. The intended mother, as I mentioned elsewhere, was fertile but did not want to have a child as she was scared of pregnancy and childbirth. Shama was quite distressed with this fact but chose to laugh it out. She expressed her distress by questioning why they had to choose her of all the surrogate mothers listed in the file. She knew that the intended parents would never tell the child about the surrogacy. She said, "I would be very happy if they would want to keep in contact with me, but they are not of that sort". The mothers in the clinic were not allowed to see the baby. Shama said, "Even when the surrogate mother is in the operation theatre having her stomach cut open, they will walk away with the child". She said she will request to see the child as she is confident of having the strength to give the child after seeing the baby. "I will argue that I can see the child and give it away, so show me the child". She emphasized, "I am the mother of the child, I have carried the baby. They may have given the genetic material but I have given the blood to the child". "Whatever I consume the child also eats, I'm carrying the baby for nine months, so at least this will have some effect on me. I will feel bad while giving the child away, but I am doing this for the sake of my children". Being a surrogate agent, she is supposed to be present along with the woman she recommended as surrogate mother during the birth and it is her responsibility to make sure that the surrogate mother gives away the baby amicably otherwise she would not get her commission. She said; "I warn them before they come here to enroll as surrogate mother, they are paying you for this, so you have to give the child, the child belongs to them, if you say you can do this, only then come here for surrogacy". She has experienced several women who do not want to give the child away at birth but it her duty as a surrogate agent to be present and advice the surrogate mothers. As a surrogate agent, she scans through the low-income localities for very poor households with serious financial problems and persuades them into surrogacy. She said washing clothes or doing housework is the main reason for miscarriage and she warns the surrogate mothers not to do any housework and checks upon them every other day as it is a responsibility the clinic has given her as a surrogacy agent.

The relationship between Shama and the intended parent had ended ever since the intended mother had informed her that she is doing this only because she was scared of pregnancy and childbirth. She didn't want to pick up her calls, so she told the doctor not to share her number with the intended parents and also did not want their phone number. During the first meeting, they also advised her to eat well and take care of the baby. When I asked her what if the child wants to know about her, she said, "That is their problem, not mine. They should tell the child, but I don't think they will ever tell the child". She further expressed very evidently frustrated, "I don't want them to feel obligated towards me, I am not doing them any favour. What they are doing is wrong, they should not opt for surrogacy when she can have a child by themselves. They are rich, they don't have the time for all this". But she also justified saying that some women can die of fear, so it's better she doesn't carry a child" and sympathized with the intended father saying "what can the man do, if the wife says 'no' and if he wants a child". I repeatedly asked her in several ways to give me her contact details but she told me to ask the doctor, and if he says she will give it to me. Hence this was a one time in-depth interview and not like the other surrogate mothers whom I followed throughout their pregnancy.

Conclusion

Surrogate mothers in India have no rights over the child, no legal or psychological support, they receive a lesser share of the total surrogacy payments, submit to unfair payment patterns, with no additional payment for miscarriages, are not safeguarded with medical/life insurance and some clinics make it mandatory that they remain in surrogate homes away from their families. Excessive embryos are implanted and desired "selective reduction" in utero is done for the preferred number of babies without taking into consideration the preference of the surrogate mother as she has to sign off all rights on medical interventions. Some clinics did not allow the surrogate mother to see the child(ren), while other clinics expected the surrogate mother to bond with the child and then they are abruptly separated from the child (ren) without a plan for any future relationship with the parents/baby or much psychological assistance. The surrogate mothers are not given any compensation if they experience a miscarriage at any stage of the pregnancy. They are over-fed as the payments are made according to the weight of the child. The higher the weight of the child the better, they are rewarded for this as a good performance of delivering a heavier child. Apart from the clinic, surrogate mothers also face performance pressures from their families.

Apart from the violations of human rights, expectations from the surrogate mothers on the assumption that they would be willing to do anything in return for money is inhuman. As feminists, we should begin asking ourselves, are we humane when we support surrogacy? Are we slipping from a post-fordist to a post-humanist era? Apart from the humanistic concerns, one significant aspect of surrogacy experienced as narrated by women in India is about their bonding with the child and considering it a part of their body or soul. Feminists have claimed that women do not necessarily bond with their babies, and rightly so, but would it be right to say that most women do not bond with their babies and denying them that opportunity or right over about the that babies they consider their own is equally a greater concern. There is also an anxiety amongst feminists to express maternal bonding. It is considered old school and conservative thinking to talk about maternal embodiment. Young's concept of pregnant embodiment, the embodied experience of infertility, pregnant person, relationship between the self and the embryo-fetus-forthcoming child explains this phenomenon (Young 2005). Crucial to Young's existential phenomenological understanding of the body in pregnant embodiment is also the understanding that embodiment is a mode of "*being-in-the-world*" wherein bodies are impacted by social norms that govern them (Young 2005, 9). The embodiment of the child perceived by the surrogate mothers as a part of their soul has a deeper cultural context of social norms. In this context to take away the child from the surrogate mother without even showing her the child's face or to expect the surrogate mothers to breastfeed and then part with the child forever is capitalization at its worst form. The cultural answer to this can be found in ancient Indian literature wherein both the social mother and the surrogate mother are considered very significant persons in the upbringing of the child. I have also referred to the bringing up of Krishna and Balarama by several mothers earlier in the Introduction.

References

Colen, S. 1995. Like a mother to them: Stratified reproduction and West Indian childcare workers and employers in New York. In *Conceiving the new world order: The global politics of reproduction*, ed. F. Ginsburg and R. Rapp, 78–102. Berkeley: University of California Press.

DasGupta, Sayantani, and Shamita Das DasGupta. 2014. *Globalization and transnational surrogacy in India: Outsourcing life*. Lanham: Lexington Books.

Ginsburg, F., and R. Rapp. 1995. *Conceiving the new world order: The global politics of reproduction*. Berkeley: University of California Press.

Goldfarb, James M., Cynthia Austin, Barry Peskin, Hannah Lisbona, Nina Desai, and Ricardo J. de Mola. 2000. Fifteen year's experience with an in-vitro fertilization surrogate gestational pregnancy programme. *Human Reproduction* 15: 1075–1078.

Gupta, J.A. 2006. Towards transnational feminisms some reflections and concerns in relation to the globalization of reproductive technologies. *European Journal of Women's Studies* 13 (1): 23–38.

Gupta, J. 2012. Reproductive biocrossings: Indian egg donors and surrogates in the globalized fertility market. *IJFAB: International Journal of Feminist Approaches to Bioethics* 5: 25–51.

Pande, A. 2010. Commercial surrogacy in India: Manufacturing a perfect mother-worker. *Signs* 35: 969–992.

Pande, A. 2011. Transnational commercial surrogacy in India: Gifts for global sisters? *Reproductive Biomedicine Online* 23 (5): 618–625.

Rapp, Rayna. 2011. Reproductive entanglements: Body, state, and culture in the dys/regulation of child-bearing. *Social Research* 78 (3): 693–718.

Rudrappa, S. 2015. *Discounted life: The price of global surrogacy*. New York: New York University Press.

SAMA, and S. Sharma. 2010. Can we see the baby bump. India: Surfilms.

Saravanan, S. 2013. An ethnomethodological approach to examine exploitation in the context of capacity, trust and experience of commercial surrogacy in India. *Philosophy, Ethics and Humanities in Medicine* 8: 10.

UN General Assembly. 1948. Universal Declaration of Human Rights, 10th December 1948, 217 A (III). http://www.refworld.org/docid/3ae6b3712c.html.

Young, Iris Marion. 2005. On female body experience "throwing like a girl" and other essays. *Feminism & Psychology*. https://doi.org/10.1093/0195161920.003.0001.

Chapter 6
The Postcolonial Paradox and Feminist Solidarity

Abstract Postcolonial feminists critique the universalization of women globally, for applying Western standards of emancipation on all women and describing non-western, women of colour as the "other". However, this postcolonial dilemma resulted in a postcolonial anxiety that resorted to pluralism and fragmented women into smaller identity groups. This chapter examines the postcolonial paradox and anxiety in the experiences of people involved in transnational movement from the "one-third world" accessing surrogacy in the "two-third world". This chapter observes this postcolonial paradox among the intended parents who accessed surrogacy in Gujarat, India, through interviews and participant observations. Women in India participate in surrogacy to enhance their economic conditions and as their body parts are comparatively more remunerative than men. A choice between "poverty" or surrogacy cannot be prescribed as "liberty". Intended parents had escaped the stricter surrogacy regulations in their home country in expectation of procedural ease, lesser rights for surrogate mothers (SMs), and more control over the surrogacy process in India. These motivations of intended parents are inherently exploitative. The experience of intended parents and the surrogacy markets reveal that it is neither the imperial Global North-South nor the re-oriental South-South patterns that can describe the holistic phenomenon of surrogacy in India. Surrogacy practices has also revealed the classist, sexist, racist elements and violation of women's bodies that is common in both the Global North and Global South and yet some academics and feminists fail to note these bridging factors. There is, hence, a need for global feminists to form an alliance based on a reproductive justice framework that aims to reduce forms of inequalities that transects class, race, gender, sexuality, disability, age, and immigration status in critique of practices that exploit such vulnerability.

Keywords Postcolonial paradox · Re-orientalism · Neo-colonialism
Otherness · Reproductive care chain · Cultural relativism

The postcolonial feminist critique that emerged in the 1980s–90s from the predominantly Western writings was in relation to representation and questions of location. They critiqued the universalization of women globally, for applying Western standards of emancipation on all women and describing non-Western and

S. Saravanan, *A Transnational Feminist View of Surrogacy Biomarkets in India*,
https://doi.org/10.1007/978-981-10-6869-0_6

women of colour as the "other" (Spivak 1988; Said 1978; Mohanty 1986). The aim of this criticism was to break this universalized vision of women, dismantle stereotypes, and reclaim identity by including the voice and concerns of all women with contextual narratives of race, sex, class along with its social, historical, and cultural background. However, this postcolonial dilemma between universalism and relativism has resulted in a postcolonial anxiety of fragmented pluralism that fragmented women into smaller identity groups. Generalizations between cultures turned into generalizations within cultures entangling itself with yet other forms of "otherness" and increased polarization in terms of race, class, language, and religion (Narayan 1998). Moreover, a focus on hegemonic patterns between cultures obscures the hegemonic practices within cultures such as; patriarchy, racism, classism, casteism, and sexism.

Globalization impacted the lives of women by providing more employment in export processing zones on one hand while it wiped out traditional forms of "localness" such as the handicraft industry in India. Postcolonial feminists have raised the question of the silencing and exclusion of vulnerable groups from claims of freedom, equality and rights in the context of neoliberal policies of economic growth (Dhawan 2009). Young women in transitional economies such as India, employed in export processing zones, constitute of women's bodies and sexualities that are passive, patient, and disciplined workers suited for long hours of work in confined space and under strict surveillance (Ong 1987). There is an overall resistance for the formation of trade unions within the surrogacy markets just as in the garment factories. Women in the global South, are drawn into such employment with low pay; they are undervalued, and experience unprotected employment at the end of global commodity chains in the manufacturing of fresh produce and garments. Women employed as temporary workers are easily expended and replaced by others who are desperate for work regardless of the insecure employment conditions. Neoliberal policies and the exploitation of women workers in export processing garment factories have been critiqued for their insistence on the economic trickle-down effect. A similar and yet more intrusive form of commodification of women's bodies, also closely linked with the global gradients of inequalities, is commercial surrogacy in countries like India (Hochschild 2011). While fertility control is implemented on poor women as subtle forms of coercion in India with little value placed on the lives of women, pregnancies are outsourced to these countries of the global South from Europe and America. One justification supporting surrogacy in countries like India is that women have few opportunities and find garment factory work more tedious (Rudrappa 2012). While both these gendered forms of earning options constitute women and their bodies in specific ways that are commodifying and devaluing, involving control over women's conduct, surveillance, and confinement, surrogacy additionally is a physically intrusive practice.

Mohanty (2003) has advocated for transnational global women's movement that questions both postcolonial essentialism as well as Western feminist hegemony. Feminists caution on the possible complicities of transnational feminism with imperialism. However, this alliance is one that acknowledges both local and global feminist issues and aims to address them by influencing global politics through a cross-border civil society network. This is a cross-border feminism that acknowledges the colonial past and unites with an aim to merge local struggles and the global women's movement. A recent event organized by global feminists was a resolution requesting the United Nations to include surrogacy in CEDAW as a violation of child and human rights recommending a global prohibition of the practice (Che Liberta 2017). The first transnational initiative towards this direction was the Paris Petition held on 2 February 2016, in favour of universal abolition of surrogacy followed by the Rome Petition (CADAC, CLF, CoRP 2016). I participated in Paris and Rome and experienced a transnational consciousness among feminists with mutual respect and complicity in resisting the extreme forms of universalization and determinism as a way forward.

Half of the surrogacy seekers in my study were from abroad (including Non-resident Indians). Other sources estimate around 60–90% of surrogacy that catered to foreign couples in India (Bhalla and Thapiyal 2013; NDTV 2015). *The Hindu* cited Ranjana Kamari's estimates of the percentage of intended couples from abroad as approximately 50% (*The Hindu* 2015). Dr. Sudhir Ajja from Mumbai said that between 2007 and 2013, 90% of the clients in his clinic were from the overseas of whom 40% were same-sex couples (Bhalla and Thapiyal 2013).

In my study, I observed a prominent imperialistic presence of intended parents from abroad not merely in numbers but in their power (in terms of foreign currency and its control) in the surrogacy market in India. The surrogate mothers told me recently that the market feels the absence of dollar inflow since it was prohibited recently in India. For the medical practitioners in India, this is a money-making business; money exchange that takes place solely through them, caesareans, breast milk, maintenance costs for surrogate homes, the bonus earned for nanny services, all had money flowing into the market, and although unequal, everyone was getting a share. Women in India participate in surrogacy to enhance their economic conditions, due to pressures to do something for their family, and because their body parts are comparatively more remunerative than men. Intended parents escape stricter surrogacy regulations in their home country in expectation of procedural ease, lesser rights for surrogate mothers and more control over the surrogacy process and the surrogate mothers in India. The medical practitioners are involved in the exploitation of surrogate mothers on behest and cooperation of the intended parents (refer Chap. 5). The state had been exploiting the surrogate mothers with laws having no protection whatsoever for the surrogate mothers. Yet, some medical practitioners and academic scholars insist that "as soon as you come to India, you call it exploitation". Moreover, the concern is not only that "rich infertile couples would exploit poor fertile women" but it's also about who gets excluded from this technology; it is the infertile poor women, and the increasing popularity of filiality that is having a grosser impact on their lives.

Nevertheless, it is neither the global North–South (imperialism) nor the South–South flow (re-orientalism) that can completely describe the surrogacy phenomenon in India. The pattern of surrogacy contracts can be best described as between the 'haves/two-third world' and the 'have-nots/one-third world' (Esteva and Prakash 1998). The advantage of the one-third/two-third world over terms like Western/ Third World and North/South is that it theoretically advances to intersectionality that includes various forms of global and local discrimination rather than the restricted geographical and ideological binaries. This chapter examines the postcolonial dilemma and anxiety by exploring the experiences of people involved in transnational movements by people from the 'one-third world' who access surrogacy in the Two-Third World. The chapter concludes that the ethnoscape is far more complex than the simplistic North-South explanation and that surrogacy is inherently exploitative both within national borders and in transnational contracts. Hence, feminists need to form a global alliance based on a reproductive justice framework that aims to reduce forms of exploitation that intersects class, race, gender, sexuality, disability, age, and immigration status. Patil (2013) notes the significance of intersectionality in understanding different forms of inequalities that transcends cultures and geographical regions which is similar to the reproductive justice approach. An analytic framework that is attentive to both the micro and the macro-politics of assisted reproductive technologies is needed towards transnational feminism with a reproductive justice approach. In the surrogacy bazaar, the intended parents are hence the other main consumers of the reproductive biomaterial sought through commercial surrogacy in India and what follows in this chapter is an analysis of their experiences and preferences. The following case studies narrate the experience of five intended parents (two Canadians, two Germans, and one NRI[1] originally from Gujarat). I had long-term interaction over two years with three of these parents (Caroline, John, and Mary) and Gayatri (the surrogate mother for Jignesh), allowing me to provide an in-depth narrative. All the quotes given in this chapter are taken from the case studies that I conducted between 2009 and 2012. As mentioned in the previous chapter, pseudonyms have been used to maintain confidentiality.

The Postcolonial Paradox and Anxiety

I recently presented a paper in France titled "Liberty for Whom: Reproductive Justice and Surrogacy Arrangements in India", identifying the structural and global injustices inherent in the transnational commercial surrogacy practices in India (Saravanan 2016). Soon after my presentation there was a buzz among some of the participants that surrogacy in India is "different" (read: "oriental" typically described as chaotic, exploitative, discriminative, vulnerable) while in other countries for example, Israel (read: "Western," efficient, rational, mature, organized), surrogacy is thoroughly controlled by the government where the surrogate mothers are emancipated. This explanation clearly overlooked an important element of my presentation that urged for more attention to broader structural inequalities within and between nations in the surrogacy practice throughout the world. This binary description is precisely the postcolonial feminist critique that emerged in the 1980s–90s of the predominantly Western writings on issues related to representation and location. The critique was against the binary representation of the

[1]NRIs: Non-resident Indians.

commonly phrased "Third-World Women" as the "Other", (ignorant, poor, uneducated, tradition-bound, domestic-family oriented, victimized) while presenting "Western Women" as educated, modern, having control over their own bodies and sexualities, and the freedom to make their own decisions. This critique was for inclusion of colonial voices and acknowledgement of their specific ethnic, cultural, and historic backgrounds (Tyagi 2014). Imposition of the cultural imperialism of social minorities on the social majorities through universal human rights has been another critique (Langlois 2009). However, the postcolonial dilemma between universalism and relativism resulted in an anxiety that resorted to pluralism and fragmented women into smaller identity groups. The focus, hence, shifted to cultural determinism and particularism based on cultural differences. As Mies and Shiva (1993) noted, one of the threats that ethical relativism poses is the acceptance of various forms of exploitative practices such as sex selective abortions, dowry, female genital mutilation, and caste system. Similarly, some of the writings romanticize women's agency even amidst exploitative or patriarchal discriminative practices (Rudrappa and Collins 2015; Pande 2010; Malpani 1998; Walley 1997). Postcolonial writings tend to romanticize agency, even within broader exploitative contexts, glorifying cultures in order to avoid being over-critical of the Other. Such writings venture into the dangers of justifying all forms of agency even within severe exploitative situations to reinforce subaltern experiences. These writings also tend to overlook the structural inequalities and the resulting discrimination and injustice, even while capitalism is emerging as one of the more urgent concern in the context of transnational commercial surrogacy.

Extreme polarization has also led to setting of different human rights standards for different parts of the world, such as sex selective abortions should be allowed in Western countries as parental family-balancing choices while it should be prohibited in Asia on discriminatory grounds because women have been exploited in these cultures (Dahl 2007; Robertson 2001). Such double standards and hypocritical assumptions not only undermine human rights commitments of equality towards all people but also fails to recognize global inequalities and the transnational movement of people seeking reproductive life changes. Sex selective abortions from a feminist perspective are sexist and a form of femicide, regardless of the preferred sex orientation, the location of the practice, or the birth order of the child. Certain practices that are outright human rights violations and restrict women's overall growth and development thus need to be challenged. There is a need to balance between the local and global, the extremely polarized approach that romanticizes women's exploitation with cultural determinism on one hand and painting everyone in the same shade on the other. Grewal and Kaplan (1994) effectively propose a transnational approach that deviate from such binaries and focus on linkages that influence every level of social existence.

Mohanty (1986), who had previously raised concerns about universalization, more recently emphasized instead on a need for forming a transnational feminist alliance and solidarity across the divisions of place, identity, class, work, and belief, and reiterates the need for intersectionality (Mohanty 2003, 2013). One of the reasons for this change in perception, as Narayan (1998) too has cautioned, is that deterministic expressions tend to divide and polarize women into more and more different groups impeding an effective transnational feminist alliance. Women are grouped into; "Western women", "Third World women", "African women", "Indian women", "Muslim women", and others and the description of cultures attributed to these groups remain fundamentally determinist. The other concern is that generalizations evident in universalization discourse also carries on to the universalistic assumptions within groups described in terms of cultural determinism, such as "Tamil women", "Punjabi women", "Maharashtrian women". Hence, the objection against universalization of Eurocentric hegemony has led to an equally hegemonic representation of "particular cultures" (Narayan 1998). Such paradigms overlook any recognition of similarities and restrict any attempts to bring women on one platform.

This explains the perspective paradox among post-colonial writers; some scholars continue to subscribe strongly to cultural relativism while others who critiqued polarity of relativism and Global-North South patterns observed and included global market patterns between the haves and have-nots without completely losing sight of the classic postcolonial imperialism as well as the importance of cultural context. Some scholars have critiqued that these observed patterns denote neo-colonialism and not post-colonialism (Huggan 1997). The critique was that postcolonial studies are unable to define which parts of the world are affected, and which are obsessed with relativistic approaches. The recent critique by post-colonial writers draws attention to the significance of intersectionality as an effort in understanding the limitation of polarized views (Mohanty 2003). Postcolonial scholars have also recognized that the global market scape and its outsourcing patterns is not limited to movement between the global North and South but also includes nodes of inequalities spread within continents and within nations (Mohanty 2003; Narayan 1998). Neocolonialism is understood as the control of former colonizers over the former colonies through political and economic institutions (Rao 2000). Neocolonialism is an integral part of the postcolonial approach and does not make the latter redundant. However, the problem of representation and the paradox within postcolonial scholars and theories continues. Although post-colonial approaches provide an analytical method, it is limited in providing a solution. Authors have drawn on neocolonialism along with reproductive justice towards an outcome-based analysis on cross-border reproductive care as discussed in the conclusion of this chapter (Fulfer 2017). Intersectionality as an analytical approach is widely applied by postcolonial writers as well as scholars using neo-colonial approach.

In the Indian surrogacy bazaar, the intended parents are the other main consumers of the reproductive biomaterial sought through commercial surrogacy and what follows in this chapter is the description of their experiences and preferences examining the postcolonial/neocolonial approach. The following case studies narrate the experience of two intended parents; from Canada (where surrogacy is legal) two from Germany (where surrogacy is prohibited) and one non-resident Indian (NRI). As mentioned earlier in Chap. 5, pseudonyms have been used to maintain the confidentiality of participants.

1. Caroline: Canada

I first met Caroline at a baby shower function that was held at one of the surrogate homes in a small town in Gujarat. She agreed to participate in my study and asked me to meet her again the next day at the hotel she was staying in. As I arrived the next day at the largest hotel in this small town (a three-star hotel), she was on the phone talking to the Canadian Embassy officials in Mumbai. I waited at the cafeteria and she joined me an hour later, excusing herself that it was her husband's birthday and she had been speaking to him. Her husband's sister would arrive soon from Canada to help her take the children back. As we ordered for some tea, she started speaking about the food in India. "I get really tired of the food, I'm just dying to go back. It's not just that the food is spicy but it's so kind of heavy, rich, lots of ghee, butter, spices. I have started avoiding eating food and I'm also non-veg (etarian) so I'm craving, oh gosh! yesterday I couldn't stop dreaming about steak. It's not just enough that there should be food and shelter, as one gets old it becomes more and more specific about what kind of food and shelter one needs."

The surrogate mothers had told me that they were being over-fed in the surrogate homes in anticipation of bigger children. I spoke to her about my experience with too much food being served in lunch boxes to the surrogate mothers. I told her that I was getting the same lunch box that the surrogate mothers were eating and the food was far too much. The same tiffin provider who supplied lunch boxes to the three surrogate homes also provided food to the hostel I was staying in. I told her "I have asked them to reduce the tiffin to one-fourth as I possibly cannot eat that much food twice a day (6 chapathis, a box of rice, dal (lentils) vegetables and yoghurt). Moreover, the food was not of a very good quality and so monotonous that I am fed up eating that although I had been here only for less than a month." Caroline told me it was not her business to find out what the surrogate mother actually wanted. She felt this was the duty of the doctor as she is paying them to do this.

Caroline perceived her food and lifestyle habits very specific and was uncomfortable with any changes to this pattern. However, she preferred to disturb the surrogate mother's life by keeping her away from her family members including a separation from their little children and making changes in her food and lifestyle pattern during the pregnancy. She herself was a full-time professional and her husband owned a computer firm in Canada. Desperate for a child she had undergone several rounds of IVF, and explained that she was on the verge of suicide. She first adopted a girl child from another South-East Asian country. However, she wasn't satisfied with one child as all her siblings had several children and she felt incomplete without many children so she thought of coming to India for surrogacy. Her husband was not in favour of this but she convinced him. She had used her friend's oocyte and her husband's sperms for this process. "My husband is a German and I wanted a boy child with German phenotype. Now my boy has those typical German features".

She told me that she chose this particular clinic because it did not charge any upfront payment until the baby is handed over. She was satisfied that she (the surrogate mother) was paid only nominal instalments during the pregnancy. The surrogate mothers in this clinic were paid (Euro 25) per month and (€5500) in the end. Caroline chose this particular clinic for easy payment method also because smaller instalments was an incentive for the surrogate mothers to take care of the child(ren) in her womb.

"One of the things that made me come to this clinic was the way the payment scheme works. While some of the other doctors ask for a lot of money upfront, this clinic takes money for the IVF treatment and only a nominal payment is made to the surrogate mother, but you don't actually pay until the very end…it's a good incentive for her (the SM) to keep the baby and not do much work so she doesn't miscarry. She (the SM) doesn't really get compensated very much until she carries the baby to the end".

The cumbersome paper work required by some other clinics was another factor why she preferred this particular clinic: "The other doctors ask for a very detailed history, but this doctor is not interested in all that. With her (the doctor) it's a leap of faith, a trust and although it's a risk from the perspective of the intended parents it's worth it." In general, she chose India because the surrogate mothers had less rights over the baby, "although it is legal in my country, the process is very complex and much more expensive than [in] India. The law expects surrogate mothers in India to sign over all rights to the baby even before the surrogacy begins, which is a big relief". Caroline also told me, "I am happy that she (the surrogate mother) was monitored throughout the pregnancy, her food and everyday life during the pregnancy was taken care of".

The surrogate mother, Ujwala, gave birth to twins (one girl and one boy) and while the passport application process took a while, Caroline wanted the surrogate mother to look after the babies. While Caroline stayed in one of the hotel rooms, Ujwala and the children stayed in the adjoining room. Ujwala was asked to breastfeed the children directly while she was in the hotel with them; Caroline had also appointed another helper to assist Ujwala with the children. She was upset that the children were responding more to Ujwala than to her own voice: "When I say something, they do not respond to me, but when Ujwala says something in her pitched tone, they immediately respond to her. They seem to be more comfortable with her". I also observed that she hardly went to the adjacent room where Ujwala and the children were staying, except to see if they need anything or to accompany when the children had to be taken to the clinic for check-up. In this case, the interaction was mainly between Caroline and Ujwala. The intended father had not interacted much with Ujwala, or her husband, nor did Caroline interact with Ujwala's husband. Ujwala did not know whose gametes she had carried in her womb. She did not know that it was not Caroline but her friend who was the egg donor. During one conversation she told me, "I carried the babies but they are not mine, they belong (genetically) to this couple; the girl looks like the father and the boy looks like the mother". I wonder if knowing that the babies she carried were not genetically Caroline's would have changed Ujwala's experience. In the next few weeks I became a regular visitor to the hotel also because Caroline and Shyamala (intended mother 3) were trying to recruit nannies to be taken along with them to Canada from India and I became involved as their translator in this process.

I accompanied Caroline and the children to the clinic the following week as she had fixed an appointment for circumcision which she wanted to cancel because she was not sure about the doctor's competency in doing it, to which Dr. Hutesh (Dr. Nisha's husband) gladly obliged, as he had found it very difficult to find a doctor who would agree to do it.

While the nurse was weighing the babies, Caroline started talking with Dr. Hutesh.

Caroline: "One thing, Ujwala is staying with us and she will stay there until Lukaz leaves because we have two rooms in the hotel, then we go back to one room. So, does she (Ujwala) go back to the surrogate home at the clinic so that she can continue breastfeeding? Right now she is staying in the hotel but as of Sunday she can come back."
Harnish: What do you want to do?
Caroline: As of Sunday, we want her to return back to the hospital.
Harnish: Fine.
Caroline: the other thing is she's (Ujwala's) been breastfeeding for us since 21 July so we want to give her some extra money.
Harnish: We have already paid her for her services extra because she had twins and now you want to pay her extra for helping with breastfeeding?
Caroline: 300 USD extra. We would like to give that money to you so that…
Harnish: She might spend this money here and there. This is Rs 15,000 and I can't tell her where to spend this money. We'll ask her, the purpose was for a house and she should use it for that purpose.
Caroline: I want to additionally put USD 1000 in her account for her use.
Hutesh: Any money you want to give her, you will have to go through us, there is no other way. What I suggest is that you transfer the money to my account and I will transfer it into her account.
Caroline: Okay, I have an idea. We can say Ujwala you have been breastfeeding so we open an account for you and your son with USD 1000.
Harnish: My suggestion is to give it some time. If she is short of money then we do it now, otherwise we wait, because these people spend the money here and there so we need to be exactly sure and we need to deposit the money in the bank.

The first question turned into a suggestion, and then into a command, and usually whatever the intended parents asked for is abided by because they are the buyers. Ujwala was staying in the hotel room with the children and Caroline wanted her to return not to her house but to the surrogate home and continue breastfeeding using the pump to which Dr. Hutesh agreed without considering the opinion of the surrogate mother. As buyers, the intended parents have full control on the surrogate mother, her body, her womb, her breast milk, and her services as a nanny for as long as they want to be served.

Harnish also did not ask Ujwala, if she wanted more money or whether her desire to buy a house was fulfilled by the money that Caroline was paying her. In fact, Ujwala fell short of money (Rs. 100,000 approximately 2000 CAD) and she had to repeat the surrogacy process. For Harnish, it was more profitable if Ujwala came back for another surrogacy than if her purpose was solved in one round of surrogacy. Caroline went back to Canada thinking she did her best to bring a poor household out of poverty, and Ujwala repeated the surrogacy to buy the house she desired.

When the time to say goodbye came closer, Ujwala said, "I am happy to have given life to these children but (I) have to give them (the children) away as a gift (to this couple) though my heart is hurting (she shudders as she says this), these children are part of my life but the deal (the contract) was made right at the beginning and I have to keep it up by giving them away". The baby starts to cry during this conversation and she puts him on her shoulder and with gentle pats tries to quiet him. Caroline on being asked about bonding between the surrogate mother and the child(ren) and trauma after the relinquishment said, "I'm not sure if the surrogates or doctor at the clinic themselves would ever use the phrase 'give the baby away'" completely unaware that Ujwala had used precisely these words "give away" just there in her presence. "These words are usually introduced by the media and the people who interview them," she continued, "but from the very beginning the doctors try to counsel the surrogates in a way that makes the surrogate aware that the baby(s) are not theirs to 'give away' and that they result from embryos belonging to the biological parents." The surrogate mothers told me that this is what the intended parents are repeatedly told and made to believe by the doctors.

After Caroline left with the babies to Canada, the girl child had somewhat settled to the changes but the boy child was very restless: "It can be very difficult to get him to stop crying…when the crying really escalates, he becomes apoplectic." She complained about not having enough assistance, she had not found a nanny, and was struggling with the children for several months until she found one (a young Filipino woman) who was dedicated and managed well with the children. Radical feminists and behavioural scientists have observed and critiqued the exploitation of women of colour as nannies for relieving the household and child care burden of the white/elite women (Fudge 2011; Hochschild 2000; Raymond 1993). Moreover, the impact of this practice on the children is completely trivialized by the doctors and similarly by the intended parents.

Regarding impact of relinquishment on the surrogate mothers after bonding and breastfeeding for several months, Caroline said, "I am told that some doctors recommend that the surrogate should not spend too much time with the baby(s) after they are born to help minimize the pain of separation. In fact, our reasons for inviting her to spend so much time with us were really to help us tend to our babies better through provision of breast milk and extra baby-minding. As she was obliging, our reasons could be considered selfish by some and while we also had an interest in getting to know her better and developing a relationship, sometimes I wonder if we did her a disservice as she was far less interested and attached to them immediately after they were born."

After returning to Canada, Caroline tried to maintain contact with Ujwala but there were constraints. One of these was that they didn't share a common language and a further limitation was because Caroline did not want to share her contact details with Ujwala. I am not sure whether this was another instruction given to her by the clinic. Caroline knew Ujwala's contact number but had never made an attempt to visit her house. On knowing about my next visit to Gujarat, she told me she will send me photos of the children through email. "I will provide a series of photos so that you can share it all with Ujwala. We would be most grateful if you could meet her and go through the pictures. Hopefully, she can find a way to print her favourite photos." Caroline trusted me to take the photos of the twins to Ujwala when I planned to visit her house. Such a fear, according to Ujwala was created by the clinic as the clinic portrays the surrogate mothers and their family members as being "money-grabbers" and "blackmailers".

Caroline did not want to do anything that would in any way counter the clinic rules. One of the rules was that all financial transactions between the intended parents and the surrogate mothers were to be made through the clinic. Caroline had deposited money to a trust fund for Ujwala's son which he could withdraw after turning 18 years of age. Caroline had funded a nursing course before-hand in which Ujwala enrolled herself soon after they left. As a participant observant, I was at the clinic when she expressed her desire to pay Ujwala more money as she got to know that her requirement to buy a house had not been fulfilled, but the doctor told her not to do so as the other surrogate mothers will start asking for more money too; "we don't want to spoil the surrogate mothers, do you know what I mean". For the doctor, it would mean another surrogacy and more business if Ujwala repeated the process. Ujwala completed her nursing course and repeated another surrogacy. She lived in a four-walled room with her husband and child with no toilet or bathroom and wished to buy something similar. Caroline had got to know from the clinic that Ujwala wanted to do another surrogacy and told me that she was unhappy about this; she felt it was a special feeling she shared with her that she wasn't happy if Ujwala shared it with any other women. She told me to tell Ujwala that "we miss her and think about her often (probably every day). Please tell her that she will forever be in our hearts and minds. We wish there was an easy way to make contact and see her more frequently. We tried to call her at Christmas time but we think Subash's[2] father ended up answering the phone. He explained that she wasn't at home." (Photo 6.1).

However genuine Caroline's motivation may seem, it was inherently exploitative because although surrogacy is legal in Canada, she chose India for surrogacy because women had lesser rights over their body and the child(ren), over her home country where women were protected by law, hence, had better decision-making rights. These motivations subscribe to the notions of the Other by exploiting the vulnerability of these women. She preferred a child(ren) with her husband's (German) phenotype despite having adopted a child from Vietnam. Several feminists have described this as a patriarchal hegemony and have strongly critiqued this sort of preferential phenotypes as a form of racism (Rothman 2000; Berkowitz and Snyder 1998; Raymond 1993). The terminology she used for describing the woman who borne the children was precisely what the doctor was using, the surrogate. The

[2]Subash is Ujwala's son.

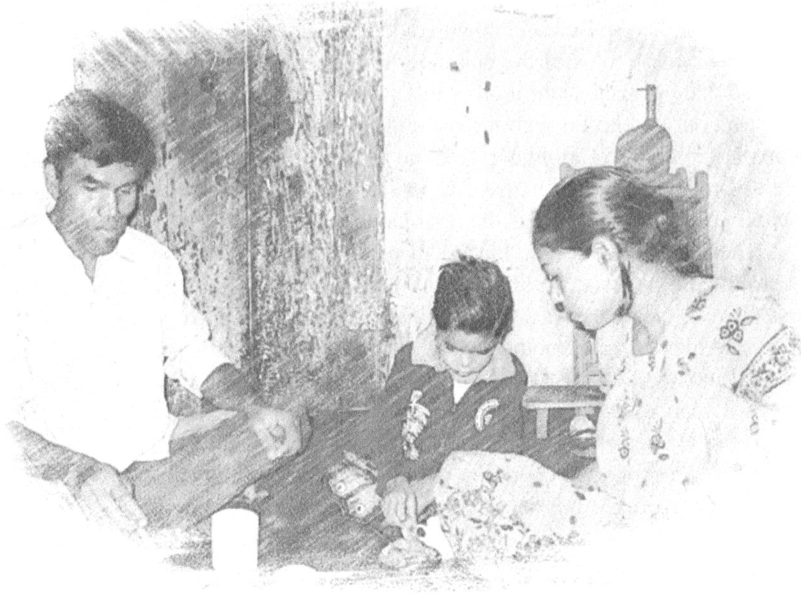

Photo 6.1 Ujwala at her home with her husband and son (Photo taken by the author, 15 November 2009)

other rhetoric that emerged from this case is the significance of filiation and genetic ties with the forthcoming child that is given precedence over carrying the birth mother. The payment for the gamete and for the procedure determined the intended mothers' ownership over the expectant child and the surrogate mother's body. These are indications of a different form of recolonization over biomaterial (body, gamete, embryos, and children).

Like other dimensions of citizenship, "biological citizenship" is undergoing transformation and re-territorializing itself along national, local, and transnational dimensions in the context of transnational surrogacy. "Biological citizenship" descriptively encompasses all those citizenship projects that have linked their conceptions of citizens to beliefs about the biological existence of human beings, as individuals, as families and lineages, as communities, as population and races, and as a species (Rose and Novas 2003). The payment for the gamete also determined her ownership over the expectant child and the surrogate mother's body. These are indications of a different form of recolonization over biomaterial (body, gamete, embryos, and children). This is also a flexible form of biological citizenship because people circumvent the state by accessing services that is legally banned in the source countries. As mentioned by Whittaker and Leng, "Previously dependent upon one's status as a citizen of a particular nation-state, now access to health care is increasingly arbitrated by one's economic, consumer or insurance status, mobility and biological status. The notion of rights to health as a universal claim is being rearticulated within a global market" (Whittaker and Leng 2016: 13). Another important aspect in India about health care exclusion, is the lack of health care services for local infertile couples versus services provided for travellers.

2. Shyamala: Canada

Caroline introduced me to Shyamala and Suresh also from Canada who had babies born through surrogacy and were staying in the same hotel for one month. Suresh is a computer engineer by profession and he runs an IT company in Canada and Shyamala is a housewife. As we approached their room, Shyamala came out of the room with a baby in her hand, she had been bathing and massaging the children along with Supriya (the surrogate mother). She explained, "We have two rooms, I keep the babies with the nannies (the surrogate mother and an extra appointed nanny were in the other room with the babies) and I am in between (the two rooms). I have a girl and a boy". Suresh is Tamil and Shyamala is Sinhalese, both originally from Sri Lanka now settled in Canada. She introduced her elder daughter (Gautami) born naturally to the couple. She's "my angel, my big baby girl", she said. She was about five years old, sitting in the middle of a large bed, only occasionally glancing at us between her Wii games, while Suresh was watching a cricket match.

Shyamala began narrating why they chose surrogacy and why India. "After my first one, we had a series of infertility problems, initially we thought one child, that's enough. But then we wanted more children mainly because we wanted our elder daughter to have siblings when she grows up. All her cousins are big and we wanted her to have siblings that she (Gautami) always wanted. We had a series of IVF and all that and we were getting old too. We spent a lot of money trying to get pregnant in Canada, but the failure was unexplained. I got pregnant a couple of times but this is actually a very good programme. We didn't want to grow too old and waste more time and one of our friends suggested surrogacy. We didn't think about this before. My friend had watched a documentary about the clinic on TV and she recommended me to do this as she was very impressed with the whole thing and she wanted me to try it out."

Shyamala was hesitant about surrogacy in the beginning but then she did some online research on this clinic and the more and more she researched, she was dragged into it. She agreed there are possibilities of exploitation. While she was contacting clinics in India she was doubtful about certain clinics, but this clinic according to her, takes care of all the surrogate mothers very well so she felt she should go to India and give it a try. She stated: "I read newspaper articles on this clinic, saw some documentary films and scanned through the clinic website." Although she had experienced some difficulties, "it was kind of hard, the entire process, waiting for the results and the ups and downs but in the end we are blessed with these babies."

"In Canada, they don't really encourage surrogacy but if you do it outside Canada there's no problem of taking the babies back," Suresh added, "they are totally against doing surrogacy for money in Canada." Shyamala said, "If you have a friend, or a family member who is really nice to you and would do it for free then you can reimburse the medical charge." According to them, "You can give them the cost of living, housing, schooling, one can pay towards a bill, but lumpsum money is not acceptable. But if it happens outside the borders and the child is proven to be genetically connected to the intended parents, then it is considered as children born for Canadians outside the country. If Canadian diplomats have children born here, the same pattern is used for surrogacy."

Shyamala was very satisfied with the surrogate mothers. "The mothers are very good. I treat my surrogate mother (Supriya) as my sister, she's there I can at any time leave the babies with her and go, no problem at all. She's giving breast milk to Ananya (the baby girl) because she is very small so she needs some breast milk. We give both the children formula too, this guy (the baby boy) is getting breast milk once per day just to keep his immune system going." She mentioned the previous week when Supriya's son was sick, "I offered her to go back home be with your son and come back (when he feels better). After initially agreeing that she would briefly go home and visit her son, she decided against it. My mom and my sister both have come to my house and they told me not to worry, she said. He's taken care of, I talked to him on the phone so I'm not going." According to Shyamala, "She (Supriya) doesn't want to leave the babies; she wants to make sure that the babies get breast milk and she's very dedicated. So she's really nice, I have a very good relationship with her, she's very sweet. I think majority of the surrogates are very kind and nice and dedicated surrogate mothers. One or two bad elements. Ya, we have in every community, but these people are nice."

Suresh says, "We have started a trust fund for that boy (Supriya's son)." Shyamala and Suresh always wanted to know "why she is going though this (surrogacy)" but did not get an opportunity to ask her during the pregnancy. Suresh said, "This time when we met her, my wife asked her and 9 out of 10 times she stressed on her son being educated. She wants her son's life to be different from hers. So we are going to set up a trust fund over and above the fees of the surrogacy, so it will be a foundation for this kid to get a better future." Shyamala added "While these babies are growing with us we want good education for her son too. We'll keep in touch, we'll be sending their photographs, we'll be always corresponding with her. She's going to be part of our family". But language would be a hurdle. Later when I went to meet Supriya along with Shymala I realized that they could hardly converse with each other. It was the other nanny from Nepal who could speak better English and was playing the role of a translator. Supriya seemingly had expressed her desire to keep contact with the children to the couple. Shyamala says, "Supriya will have to learn English as my children will never speak Gujarati or Hindi." Suresh suggested an English crash course that she should take if she wants to speak to the children. "I think we should enrol her in an English course. She speaks little English now because she has been with us for a while (almost a month)." Suresh said, "Our elder daughter, Gautami is quite open to the whole process too." Shyamala added, "She likes her (Supriya) too, the other day she was telling me mom, I wish we can take Supriya back to Canada with us. Even Supriya tells me if it wasn't for her son, she would leave her husband and come to Canada with me."

It was lunch time and Shyamala told, "You can see, I will go to her room now and she can order whatever she wants from the menu card. She's kind of tired now eating the menu, so we are trying to get some food from outside. So it's like that, anytime she wants anything she can tell me. First time we went to the Canadian Embassy, Mumbai, she came with us. That was her first experience in a flight, she was so excited, we then bought her some clothes. She's really happy. Her son wanted a TV so we went and bought a TV. They are going to deliver it (to their house) today. She called her husband and her husband is excited too. Our bonding is really nice. It's a win-win situation right, we have our babies, she gave us our babies and in return she got something that she will not get in her lifetime. Because she works as a peon in a school, her husband is an electrician. She's from a small village. Her whole house is not even half of this room and she has the bare minimum necessities. I think majority of the women are like that. So this money and gifts makes a huge difference in their lives. So both sides of the parties win in this."

Suresh asked me, "Did you get a chance to look at the Oprah Winfrey's Show on surrogacy? In that they have very eloquently articulated every step of the surrogacy process with an example of a couple from USA". On being asked they said, they would not like to change anything in this process and were quite happy, except for the anxiety level during the surrogacy process. According to Shyamala, it's not good to use surrogate mothers and then dump them on their own like some other clinics do. She said she had heard about intended parents who didn't allow the surrogate mother to bond with the baby: "Everybody is not the same, one can't expect it, one can't change. In India, a lot of exploitation is going on, look at what they do to the beggars, all these poor people. They buy their kidneys, their organs for some rupees."

Both Shyamala and Caroline were from Canada and spent lot of time together in the hotel and had similar opinions. Both insisted that this particular clinic was the best in India. There was an intense level of obligation they felt towards the clinic. They were probably completely oblivious to the problems the surrogate mothers faced within the surrogate home, and the problems her family and the surrogate mother's child(ren) would have faced with their mother away for one year. Supriya was torn between the responsibilities towards her sick child at home and the twins she had just borne and to breastfeed them. The surrogate mothers were unlikely to have shared this with the intended parents, given the implicit goodwill relationship between the intended parents and the clinic. It took me several weeks of interaction with the surrogate mothers, following them back to their homes, to their comfort zones, for them to eventually trust me enough to speak openly about their problems with the clinic, the doctor, the intended parents, and the problems they faced being in the surrogate homes. Both the Canadian intended mothers comprehended the surrogacy process mainly from an individualistic perspective and the language they spoke, the terminology they used, was very much that of the doctor's. Shyamala switched between calling Supriya, her nanny, the surrogate, surrogate mother, and sometimes even her sister but never by her first name. The lack of communication between the intended parents and the surrogate mothers was another point that came up in this case. The intended parents had a lingering question to ask the surrogate mother "why is she doing this?" Firstly, they agree that it's not easy for a woman to have a baby for someone else and give it away. Secondly, they had to wait until she had given birth to ask her this question, so it is clear that the clinic plays a gatekeeper's role in the communication between the intended parents and the surrogate mothers. At the clinic it is more business-like environment, "You want a child, I have a woman who can do it for you, you pay me for it, the remaining is our responsibility, you don't have to interfere or worry with what we do with the surrogate mothers". This was also with Caroline, when she said, it's not her business to know what the surrogate mother wants. Shymala's description of surrogacy as a "win-win situation" was exactly the words of the doctor and she practiced this diligently with the exchange of material products as gifts for the children apart from the fee also reiterating the structural inequalities and hegemonic relationship.

3. John: Germany

It was March 2010 and I had read all about John and his twins in the newspapers in India. One intended parent incidentally gave me his contact upon his consent. We spoke on the phone initially, and eventually we started regularly conversing on the phone. Finally, he invited me to Jaipur where he had been living with his twins for more than a year. I took the Rajdhani Express from Mumbai to Delhi and was quite nervous as the arrival time at Jaipur was 2 a.m. I got off at Jaipur and there was hardly anyone at the station and I thought to myself, "I am supposed to meet a German man I had never met before, in the middle of the night at Jaipur railway station! and he's nowhere to be seen." After a while I saw a short and stout man walking down the platform with his mobile to his ear, and by then my phone had started ringing. It was John, he asked me "Are you the one with golden spectacles, a green dress sitting near the station master's cabin?", I said, "Yes" and I asked him, "Are you the one with white shirt walking towards me speaking on your phone?" and both of us waved out at each other. After shaking hands, we walked out of the railway station together towards a nearby hotel where he had booked a room for me. He had insisted on booking a place in a hotel as he was more familiar with the city. On reaching the hotel, he wanted to make sure that the room they had provided for me was ok. On seeing the location of the room, he lost his temper and started shouting at the reception staff, "Have you lost your senses? How can you give a room next to the reception to a single woman? There will be hotel staff moving around at night and it is unsafe for her." He stormed out of the hotel with my suitcase and I followed him hastily with my handbag. "These idiots have no sense", he muttered. At this point I was very tired and all I wanted was to be inside a hotel room, sleeping, safely away from this angry man and any presumably abusive hotel staff. I have been quite an independent person and am certainly not accustomed to a man being overly protective about my safety. After 500 m, he walked into another hotel and this time, thankfully, he was satisfied as they had given me a room on the first floor. He walked into my room, examined it, wished me good night, and told me to have breakfast and be ready by 9 a.m. the next morning. I looked at my watch and it was 3.30 a.m. in the morning. By then I knew he was quite a personality and I was heading for an adventurous day ahead.

John, a German father was living with his twins (two boys less than two years old) in Jaipur, a city in Western India, after having been involved in surrogacy in Gujarat. Surrogacy practice is illegal in Germany, hence, the children were not accepted as German citizens. He was desperate to have children and took the risk of coming to a clinic in Gujarat, as he had been assured by the clinic that many such couples had managed to leave India (by evading the law) with fake passports and birth certificates and it would be a fairly easy procedure. After the birth of the children, he had left the country (India) to go to Eastern Europe with the help of illegal birth certificates and passports for his children. He was questioned at one airport immigration control and it was then, when the Indian Embassy enquired about the passports of his children, that the German Embassy was contacted. It soon became evident to the embassies that he had been involved in a practice considered illegal by the German government and had acquired illegal documents (passports and birth certificates) from India. They were illegal because the parent's name given on the passports and birth certificate was not according to the Indian law. The clinic had managed to provide him with birth certificates with his name as the father and the surrogate mother's name as the mother of the children which is not legal in either country. He was asked to return to India along with the children and surrender their passports. Since then, he had been involved in a legal ordeal with the government of India and the German government. According to the Indian government the children were German, but according to the German government the children were Indian as they considered the surrogate mother as the birth mother. He had been living in a luxurious service apartment along with the children and his mother (the children's grandmother).

The phone in my room rang at 8 a.m. next morning, it was John asking me if I was up and ready. I woke up and had my breakfast and the reception called me sharp at 9 a.m. about a visitor waiting for me. I knew who it was and when I went down I was pleasantly surprised to see John with the twins. The children had partially Indian phenotypes, black hair and brown eyes, as he informed me in our earlier conversations that he had used Indian oocytes and his sperms for the surrogacy. He told me, "I never take them anywhere out; this was their first outing since last few months". Later during the day, I realized the only place he took his children to was the terrace garden within the service apartment building. As service apartments are usually frequented only by people on business trips, this terrace garden was not frequented by any other children. In other words, to my utter dismay, these children had been kept virtually under house arrest and had not interacted with any local children or people, except for the father and the grandmother. He had an aversion for India and was not willing to take his children even to a local park.

His wife worked in another country and there was a family routine of speaking to her on a Skype conference call every day and that day I became part of this routine. We all walked towards the service apartment and as I entered the flat the grandmother of the children was very excited to meet me and began speaking in German with me. “How do you manage in India without eating bread, I bring all the flour from Germany but it doesn’t last for long?”, she said. To her, I was someone who had arrived from Germany with whom she could share what she misses about her country. He showed me the canned food she had brought from Germany and all the while he was in India, he tried his best to feed the children with food brought from Germany. Soon, we had to leave for the local registration office, and the children began crying as he started getting ready to leave the flat. One of the twins was visibly more disturbed and would not let him leave the house. John told me on our way out that *Oma*, (his mother) favoured the other twin who, according to her, had German characteristics and had a liking for German food, hence, this child clung on to him instead. The children had not experienced much of open air, green grass, or the company of other children of their age. They were in a difficult situation stranded with their father and Oma.

It was on our visit together to the registration office that it slowly sunk into me that he was eager to show me the worst of India; the system, the roads, and the people (especially since I was familiar with Germany as well). We went to the registration office as he was supposed to register every three months and it was time that he extended his registration. They handed him a form which he started filling out and after a while he began muttering, irritated again, and eventually began shouting at them; “How many times do you want the same information from me again and again. I guess, next, you would also want to know what brand underwear I am wearing”, he said in English. All heads turned towards me as I was the only woman in the room, and in a conservative place like Jaipur, it was disrespectful to make such a personal reference in my presence. He sensed this discomfort but didn’t care. Later he fought with an auto driver for charging us more than the meter charges and told him, “Don’t you believe in karma, you will be born as a lizard in your next life.” Very clearly his level of frustration was at its peak. I asked him on our way back if he would ever want himself and his children to be Indian citizens, if he just happened to live in India for 18 years and his answer was a clear “no”, “never”. Ironically, he was fighting a legal battle with the Indian government for obtaining Indian citizenship for his children. People who circumvent the state by accessing services that are legally banned in the source countries generally claim biological citizenship in both countries.

In 2010, the children were given permission to exit India as the German Embassy gave him some sort of visa permit to enter Germany along with his children. After reaching Germany, he told me; “We still have difficulties to realize that we are finally at home. Maybe it is still only a dream. But the children are enjoying the different emotions and way of life around them. During the first week, every day both of them ran after waking up in the morning to the window to check if all the green trees and their garden are still there. And they loved the German bread instantly”.

He was overall annoyed with the doctor in the Indian clinic, who had initially assured him of leaving the country with ease. He said, “They are crooks; they were insisting that I send my children to an orphanage as the only legal way out.” According to him, the clinic had also threatened the surrogate mother to admit that she, along with John, was responsible for preparing the illegal documents and she had to be in hiding for a while. In general, he was bitter about what he considered the ineffectiveness and selfishness of the Indian governance which created hurdles for him and this would affect the children as well who, during that phase, interacted only with their father and Oma. He had a hatred for this country (India) for having put himself and his children through this ordeal. "It should not be forgotten (how) the Indian government handled the problem. India is enjoying the benefits of foreign-exchange resulting from surrogacy. There are millions of dollars, maybe billions. But the government's behavior is typically Indian. The government is blaming unsuspecting couples for their own incompetence and their lack of responsibility is pushed off on the couples. Currently another German couple and their newborn baby are caught in the trap. They have done their biggest mistake: they have gone to India to be a family!"

His motivation to travel to India in expectation of procedural ease based on corrupt reassurance by the clinic also indicates the notion of the Other. This became even more evident when he was unable to use this corruption to his benefit, and the Indian government imposed their laws and rules upon him, which according to him they shouldn't have because they are getting revenue, just like the clinic. He then began cursing and hating this country for precisely the same reasons that attracted him to this country initially. He was genuinely caring for his children and remained with them in India for two years. Just like baby Manjhi, the grandmother got involved (details of Baby Manjhi case in Chap. 2). But he had put his children through a tough time, which would probably affect them for the rest of their lives. Moreover, he was also protecting the children from the Indian food, pollution, and people. He clearly wanted his precious children from India but hated the country. This hatred could have also developed because of his two years' wait in the country.

The German Embassy eventually provided visas to the children on the condition that there would be an adoption. However, the surrogate mother had refused to give the children in adoption to the couple. The German side of justification for providing the visas for the children was that the German public policy was not violated as the children were born through surrogacy abroad, surrogacy per se is not a violation of human dignity. The judgement draws comparison to adoption especially if the surrogate mother has no genetic ties to the child, saying it is in the best interest of the child that the couple are the legal parents especially if one intended parent was also the child's biological father. The significance of filiation is a significant determining factor for biological citizenship in this case study, as well as in the earlier case on Canada. Countries like Germany, that allow their citizens to access fertility services in countries such as India, are criticized not only for abdicating their responsibility for meeting the needs of their citizens but also for exhibiting a willingness to use the vulnerability of impoverished citizens in less affluent countries. The destination countries too have been blamed for viewing the trade merely as an opportunity for economic growth and thereby permitting objectification of their citizens. The moral reasons for which it is forbidden in Germany should also be applied to all parts of the world and should be forbidden everywhere. Hence, the post-colonial dilemma of double standards that protect

German citizens while other citizens may be exploited is evident in this case. Again, the impact this has on the children is not taken into consideration by any of those involved in the process.

4. Mary: Germany

On my visit to a low-income settlement near the clinic to meet a surrogate mother, I saw 6 months-old twins (a girl and a boy) crawling in a house and was told that they were German children born through surrogacy. I enquired about their parents and came to know that apparently, the mother visited India only once every three months. Later, I learnt that she was on a tourist visa which expired every three months. The boy child had a whooping cough and the girl was also not doing too well. The boy was almost starving because he didn't eat much. This house was not of the surrogate mother who had borne the children but another surrogate mother from the clinic who was taking care of them as a nanny. At that point I was wondering, "what are these intended parents thinking when they come to India"? One can only imagine the state of infants who have had to change hands every three months. I had an opportunity to meet Mary the intended mother of these children on her following visit to India. Her children were born in 2009, after John was caught with illegal passports and the German government had become more alert and aware of these issues. Since her arrival, Mary had moved into a lavish apartment along with the children. When I entered the apartment, Mary was feeding, or one can say over-feeding, the children. She had brought all the baby food for them from Germany. She complained that her children had become frail and sick and Jyoti (the nanny) had not been taking good care of the children (Photo 6.2).

Mary, an intended mother of an African origin, was living in Germany and married to a German man. She had been facing infertility problems due to her uterus lining and tried several IVF trials first in Germany, then in Spain, and also at a popular clinic in Mumbai. She had been advised by the doctor in Mumbai not to waste any more time and money on IVF and to opt for surrogacy. He recommended her to go to this clinic in Gujarat for surrogacy. Her husband had not visited India since he deposited his sperm. Her surrogate mother conceived at the first attempt and gave birth to one girl and one boy in 2009. She arrived in India six weeks after delivery, meanwhile the babies were kept in an incubator for three weeks and thereafter the surrogate mother was looking after the babies in a children's hospital. Mary also informed me she had to pay double the price as the charge in the clinic was per child not per surrogacy. She had thus spent a total of 30,000 Euros on the surrogacy process in India. Her relationship with the surrogate mother also ended on a bitter note and since then she had not kept any contact with her. Initially, she wanted the surrogate mother to continue to breastfeed the babies and assist in taking care of them, so she brought the couple (the surrogate mother and her husband) to her flat. Due to some problem between Mary and the surrogate mother's husband, she made a payment and asked them to leave. Then she found another nanny (Jyoti) to look after the children. She was unhappy with Jyoti too because she felt that she was not caring for her children well as she was always on her phone. She also suspected that Jyoti was involved in prostitution. Jyoti had a room to herself in the

Photo 6.2 Author with German children born through surrogacy stranded in India (Photo taken by the author, 23rd February 2010)

apartment and had many visitors, who Mary suspected were not her friends or relatives but her clients. She told me, "I think most married Indian women are involved in prostitution and simply act as if they are 'good wives'".

She blamed John for exposing himself and thus causing her to remain in hiding as the German Embassy had become more alert. She wanted the clinic to arrange for illegal documents just as they had done for John, but they were refusing to do so. She told me that any illegal activity is possible in the clinic because they had the contacts and influence, but they refused to help her since John's case was exposed. According to Mary, "This is a poor country, everything should be possible by paying money; people's mouths can be shut". She even requested me to explore illegal ways to take the children out of India. She was delighted to inform me that "a German man had commissioned surrogacy with an unmarried girl in Mumbai. He can prove (to the German government) that he came to India, had an affair with an unmarried girl in India and had a baby with her. So there are such loopholes and hopes for us". Now there is another way out for surrogacy she said happily. A similar case was known of a gay couple in Germany who had a child born via surrogacy to an unmarried surrogate mother in California in 2010, the children were granted citizenship on the same grounds already discussed earlier in John's case.

Mary was trying to smuggle the children out to Africa, her place of origin. She blamed the German government for being selfish and the difficult law that keeps her children stranded in India; "My children are living in India. Wherever human beings live, it is their home. India says the children are not Indians and that they were only born in India and Germany says the children are Indian because they are born through surrogacy, so which country does the children come from then. All other children who were born (at) the same time (April 2009) or later have gone home except my children who were unfortunate to be born to a wicked country like Germany and heartless country like India."

Her motivation to move to India for surrogacy was similar to John's; she expected the corruption to assist her illegal exit to Germany with the children and the assurance by the clinic prompted her to believe so. She was extremely judgemental about India, the Indian system, the food, and Indian women. However, just like John, this surrogacy proved to be a difficult track, as she has been in hiding in India for almost a year. In this case, the children were the ones suffering because they had to change hands every three months as well as change residence and food patterns. This case makes me wonder whether intended parents who come to India from prohibitive countries are genuinely oblivious to these realities, or does this come from sheer irresponsibility, or is it an overpowering desperation that takes over common sense. As Carmel Shalev rightly questions in a documentary film on ARTs "Future Baby", "Does right to parenthood mean 'by whatever means'?" (Arlamovsky 2016) which may imply at the cost of the health and well-being of the children who are being brought into this world?

5. Gayatri (Surrogate mother)–Jignesh (A Non-Resident Indian (Gujarati) from USA)

I would have to narrate this case as experienced by Gayatri (the surrogate mother) because it was through her and her husband that I was briefly introduced to Jignesh. This was the other clinic, without surrogate homes. After waiting for more than a month the doctor had not introduced me to any surrogate mother. One day suddenly, I received a call from the clinic that I should come immediately if I would like to talk to a surrogate mother. I took an auto rickshaw and reached the clinic within 10 min. A nurse asked me to follow her into an in-patient room on the second floor and told the lady sitting inside to talk to me and answer my questions. I introduced myself and began asking them, the lady and her partner, about their surrogacy experience. After every question the two of them would whisper to each other. When I asked about their remuneration, their answer shocked me as it was much less compared to what I had been informed by the clinic and what other intended parents had told me was paid to the surrogate mothers. Before I could gather my wits, a nurse entered the room and summoned them downstairs for a check-up. We left the room and I was very disappointed that my meeting with them ended so abruptly. We were all together in the elevator and the husband hurriedly scribbled something on a piece of paper and cautiously put it into my hands without anyone noticing. On leaving the clinic, I opened the chit to find a mobile number written on it. The next day, I was to leave for Germany. On reaching Germany, I called this number and a lady answered the phone, it was Gayatri. She was very happy that I had called and sounded like an altogether different person from the one

I had met in the clinic environment. She was chirpy, happy, loud, bold, and articulate. On being asked about this difference, she told me "Sheela-ben,[3] there are cameras fixed in every room inside the clinic and we (she and her husband) were told that there would be a lady who could ask us questions but we should not discuss about remuneration nor give you our contact details. We were summoned as soon as you began asking questions about the remuneration. As the elevator is the only place without a camera, my husband wrote my mobile number and put it in your hands". This started my association with Gayatri and her family, that still continues till today. I was a regular visitor to their house until I was in Ahmedabad. The intended father had told them to shift their residence as he didn't like the environs she was living and he didn't want her to be carrying his children and living in a low-income settlement. She had hence lost the social support of her neighborhood. She had not informed her family members about her surrogacy involvement, so I was the only visitor to their house during that phase. Eventually, I became involved in caring for her young children when she had to visit the hospital. I connected her to other surrogate mothers I had met at the first clinic who offered her solidarity.

Jignesh and Trupti (her intended parents) were originally from Gujarat but lived in America. After several attempts to have a child, they eventually came to India for surrogacy. While Gayatri was pregnant in Gujarat, Trupti had faked a pregnancy to her family members and friends both in India and the USA by using a belly belt. After the third month of pregnancy Gayatri (the surrogate mother) was informed that she was carrying triplets. The clinic asked her opinion (for a possible selective in-utero abortion) and she left the choice to the intended parents. The next day she received a call from Jignesh that she should keep all three babies and both agreed on the condition that he would pay her extra (Rs. 50,000) for each baby. During the seventh month, Jignesh found out that the clinic was over-charging him so he contacted Gayatri again and requested her to break-off from the fertility clinic for which he would give her double the agreed amount and after giving it some thought she agreed to this condition too. The tension began when the clinic found out about this and started harassing her to come to the clinic immediately and Gayatri was worried that they would conduct a cesarean on her in order to control the monetary transaction by holding on to the babies. "I was very confused", she told me "whether I should listen to Jignesh or to the clinic". On one hand the clinic personnel began warning her, "If you disassociated from us and anything happens later we are responsible". While on the other hand, Jignesh was reassuring her by saying, "If you trust me, I will give you the money even before the delivery". He gave her the contact of a new clinic where she should go for delivery and she obliged.

On the day of delivery, in the morning, Jignesh and Trupti arrived from America. Immediately after the cesarean they took the babies to another hospital. I, along with Gayatri's husband, was waiting outside the operating theatre and Trupti was the first to emerge. She headed straight towards her husband and

[3]"Ben" means sister in Gujarati language, every woman is called with this prefix.

whispered into his ears "one is a boy". Shortly they brought the three babies out and whisked them away to a neonatal intensive care unit in another hospital. They had all fitted into one tea-tray as they were brought out and barely managed to cry; they were rather squeaking like little chicks as they were grossly pre-matured babies. Almost until four days the couple had not returned to check on Gayatri nor was there any talk about the payment. Gayatri's husband became anxious and started contacting Jignesh for the payment to which he said, "Now, take Rs. 75,000, the remaining money I will give later". After few days they received a call from him asking "how much more money should I pay you?" They were confused by his question and told him "whatever you promised us over the phone". He then began making excuses to Gayatri, "I am a very poor man; I don't have that kind of money now. Now I have only Rs. 1,00,000, the remaining I will pay when I return to America". They had no other choice but to accept what he was saying. Gayatri told me, "I feel very embarrassed firstly because he is a man and I felt uncomfortable demanding money for his children I had borne". That was the last time he ever contacted them, even seven years later there is not even a phone call from him, the telephone number and address he shared with Gayatri and her husband did not exist, and the local Indian mobile number he had given them soon became invalid. She was very keen to see the children but they were not shown to her. She was not paid extra for the babies or for shifting clinics, she was eventually paid an amount lesser than she was originally promised at the first clinic. She repeated the surrogacy process in another clinic as her requirement to buy a house had not been fulfilled. A few months later as a retrospect she added, "I carried his babies for nine months and he (the intended father) didn't even drink a glass of water in my house" (Saravanan and Ranadive 2010).

This case brings forth the significance of exploitation by people also originally from the global South. The Non-resident Indians (NRIs) returning to India for surrogacy also exploit the surrogate mothers. Hence, in this globalized world it is not only about people in the global North exploiting people in the global South, but also those from the global South taking advantage as well. In the case of Jignesh, it was not about the fear of the unknown Other, but it was a sheer business dealing between the intended parents and the surrogate mothers, bargaining about how many children she would/should carry to term, how much she wants, or how much he is willing to give for every extra baby that ended in deception. Hence, in a way these clinics act as gatekeepers but on the other hand they also protect women from such deceptions. But it is a case of choosing between the frying pan or the fire.

The Perceived 'Otherness'

Both the intended parents and the clinics were exploiting the knowledge and the system for their individual benefit. Others were in a victim mode denying any responsibility for their actions. Jignesh used his knowledge about India to manipulate and exploit the surrogate mother. Jignesh saved himself from being monetarily exploited by the medical fraternity, but many others from abroad are overcharged.

Both the intended mothers from Canada were very particular about what they wanted from the process but while they did not have much control over some elements of the surrogacy process, the other problems the surrogate mothers were facing were either obscured to them or they knew about it but were not concerned. Despite signing the consent form, Caroline wanted to control my study findings and after her participation urged that she wanted her interviews to be used only if I would write positive aspects about the clinic and the surrogacy practice in India. This prompted me to think about the bias of the participation of the two Canadian intended parents. They both were living on the same floor in the hotel, had become good friends, and participated in the study only because they wanted to put forth their voice in support of the clinic and to justify the surrogacy practice as there was growing criticism about it in India.

The Canadian intended mothers prioritized their own requirements over concerns about the effects of the mandatory stay in surrogate homes on the surrogate mother's lives, and her children. Their obligations were to the clinic and they were not willing to do or say anything that would thwart the reputation of the clinic. Caroline even went to the extent of telling me that her real name could be used only if I write in praise of the clinic and if I ever write anything against the clinic a pseudonym should be used. Both the Canadian intended parents wanted the surrogate mothers to be dedicated in her service post birth; as a nanny, breastfeeding in return for which she would get extra money, good food, clothing, travel, and gifts. This give and take defines their relationship post birth and it was the surrogate mothers who are meant to be more obliged to follow the requirements of the intended parents if they wanted a continued relationship.

The two German intended parents accepted to be participants in my study because they wanted to voice their criticism towards the Indian and German governments. Both the German couples had an aversion towards the Other and yet had come to this country for gathering the seemingly best gift of their lives (the children). They had evaded the law and swindled both the nations but to them that seemed beside the point. They came to India expecting a chaotic situation where anything illegal was possible, not knowing the whole truth. Legal court cases can be a cumbersome and exasperating process lasting for long periods of time which is the same in both India and Germany. However, they expected both countries to treat them with utmost grace and respect. The ones suffering the brunt of all this were the children along with the struggle, the suffering, and the hatred of the parents. Local food was a concern for all the intended parents. Intended parents evaded a legally controlled surrogacy in the source country to seek compliance and flexibility in India. They preferred that the surrogate mothers be monitored throughout the process, ensuring the maintenance of a strict diet and lifestyle of surrogate mothers, infant breastfeeding, the use of their labour as nannies, and preference for a bulk final payment on relinquishment of the child(ren) which reflected in the functioning of the surrogacy markets and some of the mandatory rules imposed on the surrogate mothers. This further reinforces global and structural inequalities and causes certain violations of basic human rights in the process of surrogacy in India (see Chap. 4).

Towards Feminist Solidarity and Reproductive Justice

Liberal feminists support the idea of procreative liberty and individual rights to enter into any contract that one may desire. On these grounds, harm reduction and harmonization has been presented as a solution to reduce infertility on one hand and to reduce inequalities on the other. Surrogacy is presented as a solution for infertility as those experiencing infertility face issues such as social stigma, psychological problems, physical stress of infertility treatments, and violation of bodily integrity. But, by opting for surrogacy in order to resolve this, one is inclined to put another woman through the same set of problems; social stigma, psychological challenges, violation of her bodily integrity and even more, put the surrogate mother's health, freedom, liberty, and even life at stake. Violation of another person's dignity, integrity, and economic needs hence cannot be a constitutional right. Moreover, technologies such as surrogacy reinforce inequalities as they provide a wider reproductive choice only for affluent people at the cost of the health, freedom, and life of some others (mostly the less affluent women) while designating substantial control and power in the hands of intermediate agencies. In a conference I recently attended, a comment was made that surrogacy does not create more inequalities in transitional economies than already exists. But, by overlooking the already existing structural inequalities, we ignore the effect of global capitalism on the bodies and lives of women and girls in these countries. The systematic movement of people from elite countries to transitional economies and the selected access of the infertile affluent to reproductive biomaterial using global inequalities and vulnerability of people is a sign of capitalization and globalization that is recolonizing women's bodies and labour.

Considering that surrogacy is a form of violence against women and that presently there is a clear flow of people accessing surrogacy in India because of the contextual vulnerability of women, this practice is crucial to the postcolonial discourse. As pointed out by several postcolonial theorists, there is a continuity on the pattern of rescue narratives, both in the media reports and in the experiences of intended parents who emphasize on playing a significant role in rescuing women from poverty while engaging in practices within surrogacy that are opposed to gender justice. As scholars have noted, there has been a continuity in the pattern of these rescue narratives. Postcolonial writers have critiqued how the situation, or the position of the subaltern women, has been mobilized and instrumentalized as an alibi to justify imperialistic practices. It is important to critically engage with racist Euro-American discourses while at the same time be able to critique practices within the social majority communities and societal contexts that are antagonistic to gender equality and gender justice.

In her earlier work, Mohanty (1986) focused on the Western-Third world feminist distinction in which she downplayed trans-cultural commonalities. But in her more recent writings (Mohanty 2003), she refers to one-third and two-third worlds and emphasizes on a necessity of transnational feminist solidarity against capitalism. The concern is that capitalism is taking new forms of recolonization of women's bodies and labour through existing inequalities. Capitalization has racialized and sexualized women's bodies for profits, hence, an analysis of globalization needs to include the struggles and experiences of poor women and women of colour. Although the transnational movement between the one-third to the two-third world can be described as Global Movement for Reproductive Life Changes (GMRC), these are life changes that exploit the vulnerability of people from the two-third world. Transnational feminism is not a new concept; several academics have realized the significance of such an alliance (Fernandes 2013; Mohanty 2013; Gupta 2006). Postcolonial scholars explore intersectionality as an approach that transcends geographical boundaries and cuts across various multiple forms of inequalities. Hence, there is a need for multicultural, global, and postcolonial feminists to form an alliance based on a reproductive justice framework that aims to reduce inequalities by recognizing the histories of reproductive oppression in all communities globally, changing structural power inequalities to identify and address multiple oppressions of race, class, gender, sexuality, ability, age, and immigration status. In order to understand and address inequalities there is a need to build feminist solidarity across division of place, identity, class, work, and belief. Mohanty (2003) observes that although it is difficult to build such a global feminist alliance, it has become ever so important to do so. The aim of such alliance needs to confront challenges within contexts that restrain women's overall growth and development. Tong concludes drawing on Aristotle's words that the feminist solidarity should be "partner in virtue and a friend in action" (Tong 2009: 236).

References

Arlamovsky, Maria. 2016. *Future baby*. NGF: Nikolaus Geyrhalter Filmproduktion.

Berkowitz, Jonathan M., and Jack W. Snyder. 1998. Racism and sexism in medically assisted conception. *Bioethics* 12 (1): 25–44.

Bhalla, Nita, and Mansi Thapiyal. 2013. Foreigners are flocking to India to rent wombs and grow surrogate babies. *Reuters, Business Insider*. http://www.businessinsider.com/india-surrogate-mother-industry-2013-9?IR=T. Accessed 15 Aug 2017.

CADAC, CLF, CoRP. 2016. Assisi of 2 February 2016 for the universal abolition of surrogacy. Coordination of Associations for the right to abortion and contraception, CLF Coordination Lesbienne en France, CoRP—Collectif pour le Respect de la Personne.

Che, Liberta. 2017. Bando universale, atti del convegno alla Camera. *Che Liberta*. http://www.cheliberta.it/2017/03/30/atti-incontro-maternita-al-bivio/#.WNzUzXaBf58.facebook. Accessed 14 Aug 2017.

Dahl, E. 2007. The 10 most common objections to sex selection and why they are far from conclusive: A western perspective. *Reproductive BioMedicine Online* 14: 158–161.

Dhawan, Nikita. 2009. Postkoloniale Theorie. In *Raumwissenschaften*, ed. S. Günzel, M. Castro Varela, and S. Randeria, 308–323. Frankfurt Main: Suhrkamp.

Esteva, G., and S.P. Madhu. 1998. *Grassroots post-modernism: Remaking the soil of cultures*. London: Zed.

Fernandes, L. 2013. *Transnational feminism in the United States: Knowledge, ethics, and power*. New York: New York University Press.

Fudge, J. 2011. Global care chains, employment agencies, and the conundrum of jurisdiction: Decent work for domestic workers in Canada. *Canadian Journal of Women and the Law* 23: 235–264.

Fulfer, Katy. 2017. Cross-border reproductive travel, neocolonialism, and Canadian Policy. *International Journal of Feminist Approaches to Bioethics* 10 (1): 225–247.

Grewal, Inderpal, and Caren Kaplan. 1994. Introduction: Transnational feminist practices and questions of postmodernity. In *Scattered hegemonies: Postmodernity and transnational feminist practices*, ed. Inderpal Grewal and Caren Kaplan, 1–36. Minneapolis: University of Minnesota Press.

Gupta, J.A. 2006. Towards transnational feminisms some reflections and concerns in relation to the globalization of reproductive technologies. *European Journal of Women's Studies* 13: 23–38.

Hochschild, A.R. 2000. The nanny chain. *The American Prospect* 3: 32–36.

Hochschild, Arlie. 2011. Emotional life on the market frontier. *The Annual Review of Sociology* 37: 21–33.

Huggan, Graham. 1997. The neocolonialism of postcolonialism: A cautionary note. *Links and Letters* 4: 19–24.

Langlois, A. 2009. Normative and theoretical foundations of human rights. In *Human rights: Politics and practise*, ed. M. Goodhart. Oxford: Oxford University Press.

Malpani, A. 1998. Preimplantation genetic diagnosis (PGD) of the embryo for preconceptional sex selection: Right or wrong? *Obstetrics and Gynaecology Today* 3: 486–490.

Mies, M., and V. Shiva. 1993. *Ecofeminism*. London: Zed.

Mohanty, C.T. 1986. Under western eyes: Feminist scholarship and colonial discourses. *Boundary* 2 (12): 333–358.

Mohanty, C.T. 2003. "Under western eyes" revisited: Feminist solidarity through anticapitalist struggles. *Signs* 28: 499–535.

Mohanty, C.T. 2013. Transnational feminist crossings: On neoliberalism and radical critique. *Signs* 38: 967–991.

Narayan, U. 1998. Essence of culture and a sense of history: A feminist critique of cultural essentialism. *Hypatia* 13: 86–106.

NDTV. 2015. Women selling bodies to earn livelihood: Minister Anupriya Patel on surrogacy bill. *NDTV*. http://www.ndtv.com/india-news/women-selling-bodies-to-earn-livelihood-minister-on-surrogacy-bill-1453387. Accessed 15 Aug 2017.

Ong, Aihwa. 1987. *Spirits of resistance and capitalist discipline: Factory women in Malaysia*. Albany: SUNY Press.

Pande, A. 2010. Commercial surrogacy in India: Manufacturing a perfect mother-worker. *Signs* 35: 969–992.

Patil, V. 2013. From patriarchy to intersectionality: A transnational feminist assessment of how far we've really come. *Signs* 38 (4): 847–867.

Perappadan, Bindu Shajan. 2015. A Setback for surrogacy in India. *The Hindu*. http://www.thehindu.com/opinion/op-ed/a-setback-for-surrogacy-in-india/article7927730.ece. Accessed 15 Aug 2017.

Rao, Nagesh. 2000. 'Neocolonialism' or 'globalization': Postcolonial theory and the demands of political economy. *Interdisciplinary Literary Studies* 1 (2): 165–184.

Raymond, J.G. 1993. *Women as wombs: Reproductive technologies and the battle over women's freedom*. San Francisco: Harper.

Robertson, J.A. 2001. Preconception gender selection. *American Journal of Bioethics* 1: 2–9.

Rose, N., and C. Novas. 2003. Biological citizenship. In *Blackwell companion to global anthropology*, ed. A. Ong and S. Collier. Oxford: Blackwell.

Rothman, B.K. 2000. *Recreating motherhood*. New Brunswick: Rutgers University Press.

Rudrappa, Sharmila. 2012. India's reproductive assembly line. *Contexts* 11 (2): 22–27.

Rudrappa, S., and C. Collins. 2015. Altruistic agencies and compassionate consumers: Moral framing of transnational surrogacy. *Gender & Society* 29: 937–959.

Said, Edward. 1978. *Orientalism*. New York: Pantheon.

Saravanan, S. 2016. Liberty for whom: Reproductive justice and surrogacy arrangements in India surrogacy: Situating France within the world—Representations, regulations, and experiences, 17–18 Nov, Paris.

Saravanan, S., and R.R. Ranadive. 2010. Mother anonymous (documentary). India: Frame of Mind Communications Pvt. Ltd. Mumbai and Asia and Europe in a Global Context, University of Heidelberg, Germany.

Spivak, Gayatri Chakravorty. 1988. Can the Subaltern Speak? In *Marxism and the interpretation of culture*, ed. Cary Nelson, 271–313. Champaign: University of Illinois Press.

Tong, Rosemarie. 2009. *Feminist thought: A more comprehensive introduction*. Philadelphia: Westview Press.

Tyagi, R. 2014. Understanding postcolonial feminism in relation with postcolonial and feminist theories. *International Journal of Language and Linguistics* 1: 45–50.

Walley, C. 1997. Searching for "voices": Feminism, anthropology, and the global debate over female genital operations. *Cultural Anthropology* 12: 405–438.

Whittaker, A., and C. Leng. 2016. Flexible bio-citizenship and international medical travel: Transnational mobilities for care in Asia. *International Sociology* 31 (3): 286–304.

Part III
Feminist Solidarity Towards Humanitarian Assisted Conception

Chapter 7
Transnational Feminism for Reproductive Justice

Abstract Assisted reproductive technologies have provided a wide range of choices and opportunities for people to have children using genetics of their choice (either their own genetics or others). These technologies can also be used in ways that are harmful to communities with lower access to resources and power. An analytical framework using stratified reproduction in the context of surrogacy in India reveals that some individuals gain reproductive empowerment at the cost of the health and even life of some other women based on inequalities. The reproductive rights framework is inadequate in understanding this stratification. Scholars and activists have rigorously engaged in debates and discourse through their writings regarding women's agency and their broader social empowerment from a variety of critical perspectives especially keeping in view the structural inequalities social injustice in the commercial markets of reproductive labour, babies, and bodies. In recent years, several scholars have suggested the reproductive justice framework as a way forward towards understanding and addressing such social and global injustice (Mohapatra in Ann Health Law 21(1):191–200, 2012; Roberts in Signs 34(4):783–804, 2009; Jesudason and Kimport in Front J Women Stud 34 (3):213–225, 2013; Luna and Luker in Ann Rev Law Soc Sci 9:327–352, 2013; Gupta in IJFAB Int J Fem Approaches Bioeth 5:25–51, 2012; Galpern, Presentation for SisterSong's "Let's Talk About Sex" Conference, Chicago, IL, 2007; Bailey in Hypatia 26:715–741, 2011; Fixmer-Oraiz in J Women Stud 34(3):126–163, 2013; West in Yale Law J 1394–1432, 2009; Gaard in Eth Environ 15(2):103–129, 2010; Donchin in Bioethics 24(7):323–332, 2010; Callahan and Roberts in Faculty Scholarship, Paper 1155, 1996; Dickenson in Bioethics 15(3):205–217, 2001). Building on this scholarship, I examine a possible strategic pathway to take this forward by drawing on the reproductive justice framework towards a transnational feminist solidarity that recognizes the intersectionality of structural social oppressions operating through historic systems of postcolonial and neocolonial domination.

Keywords Transnational feminism · Reproductive justice · Intersectionality

S. Saravanan, *A Transnational Feminist View of Surrogacy Biomarkets in India*,
https://doi.org/10.1007/978-981-10-6869-0_7

The global surrogacy bazaar reinforces racist, ableist, sexist, nationalist, and casteist prejudices normalized and naturalized through practices of geneticization, alienation, and commodification supported by unrestrained access to all sorts of ART services possibly available and through inadequate or ineffective laws. The surrogacy market based on the demand and supply of free-of-cost or cheap and uncomplicated wombs, hence, cannot be a solution to infertility or to the profound socio-economic inequalities. These technologies provide a wider reproductive choice for affluent people at the cost of the health, freedom and life of some others. Reproductive justice aims to reduce inequalities and not to use someone's vulnerability as a solution for infertility. All the scholars who have conducted empirical research on surrogacy in India have agreed to the exploitative and prejudiced component of the industry (Pande 2014; Vora 2009; SAMA 2012; Saravanan 2010; Deomampo 2013; Rudrappa 2015). All those who have and even those who have not researched on surrogacy have admitted to a possibility of exploitation within this industry. In Chaps. 4, 5 and 6 we have seen the extent to which this exploitation, which is embedded on structural, pronatalist, patriarch, racial, and ableist inequalities, can reach in countries like India. It has also been observed using the analytical framework of stratified reproduction that the class, race, ethnicity, gender, and place hierarchies influenced by inequalities of access to material and social resources results in differential experience and value and reward in the child-bearing process of surrogacy in India. Bringing the voices, perspectives, and experience of surrogate mothers in the centre of the discussion also broadly reveals a situation of complete control and domination over women's bodies by the medical practitioners under the guidelines of the buyer's desires during the surrogacy process. Whatever little agency women apply is limited within this broader context of domination and exploitation. As warned by Sunder Rajan (2000), these are forms of agency that is applied to maintain patriarchy, the domination. Women as surrogate mothers are rewarded with a bonus amount for being good (docile, passive) participants in this objective situation of disempowerment. They have indeed been compensated within this hierarchical social arrangement to remain at the bottom of the structure and to maintain this hierarchy. Clearly in the context of surrogacy in India, the medical practitioners as the dominant players have a well-defined collective identity; they have even organised a movement dominated by profit makers to resist the new Bill in India that aims to prohibit commercial surrogacy in India. There was no such movement encouraged to resist the dominance of the medical practitioners when the practice along with its human rights violation and extreme forms of ethical malpractices was ongoing as the practice was permitted but unregulated in India. Hence, medical practitioners and other profit makers have more to gain and less to lose by organizing protest and opposition than the surrogate mothers. Any organized resistance by women would mean eviction from the chain of hegemony, along with the possibility of any gains such as compensation and hence there is no promotion of trade unions or surrogacy organizations. As noted by April L. Cherry, "Domestically, globalization results in inequality or in increasing existing inequality because the reduction of trade barriers and investments create a greater gap in the level of opportunity afforded to those who are able to traverse transnational boundaries and those who are not able to do so" (Cherry 2014: 267).

Pande (2014) in the conclusion of her book suggests that an ideal beginning of the transnational feminist movement should be formed with a community of women with common interests which begins with defining surrogacy as "womb work", reproductive labour, similar to sex work, care work, and other intimate forms of labour. This definition is inadequate and incomplete as has been observed in my study findings; the phrase "*achha kaam*" is misunderstood as "work" and surrogacy has been a drastic step which they were compelled to take due to poverty. "Work" in this form was the last resort of earning source and defined as only one of the multiple other ways by which women experienced exploitation. Pande (2014) suggests a way forward towards a transnational feminist movement with a community with common interests, which in the case of surrogacy are veteran surrogate mothers. She explains, "Most veteran surrogate (mothers) agree that open negotiation is desirable" between the surrogate mothers and her clients (Pande 2014: 182). My understanding of veteran surrogate mothers is that they are those who have experienced surrogacy more than a couple of times and are seasoned surrogate mothers in the biomarket and in most probability, are surrogate agents coercing other women into surrogacy to earn money. Indeed, they would be eager that women continue to be exploited and coerced as they benefit from this profit-making industry. Scheper-Hughes questions are apt in this situation "Whose voices are being silenced? What unrecognized sacrifices are being made? What lies behind the rhetoric of gifts, altruism, scarcities and needs?" (Scheper-Hughes 2000: 192).

I have asked different surrogate mothers on the phone about their response to the prohibition on surrogacy, some are relieved that they do not have to prove their loyalty to their family by participating in this "*magaj mari*" (unnecessary laborious work), others are also disappointed that they will not be able to repeat it to meet their family target, the veteran surrogate mothers who were surrogate agents and earning money through this business are the most disappointed. These are some of the heard voices, but there are some unheard voices; women (first time surrogate mothers) who are being coerced by family members into surrogacy. Ask women who have been trafficked into surrogacy, or raped and forced to have children to be sold in the market whether they want surrogacy to be legalized in India. Women are also often blamed for infertility, suffer ostracism and are being subjected to various human rights violations as a result. Ask infertile women in India who have been abused or deserted by their husbands because they cannot become a surrogate mother about their opinion of surrogacy in India. These women are deemed "waste" bodies by family members and are discarded. We should ask first-time surrogate mothers who want to quit because they cannot bear the pain of the injections but prefer to tolerate the pain than to disappoint their family members who have financial expectations and who consider their bodies as the tool to get them out of poverty. The family decides how much money is a requirement for their financial security and this amount determines how many times the woman is involved in surrogacy. Even if a surrogate mother is reluctant and wants to quit on account of the unbearable pain of the injections or the psychological turmoil she experiences, she cannot withdraw at that stage because all the hopes of her family is on her shoulders. After one surrogacy, she is compelled into it again as the target has to be achieved even at the cost of her life. She is bound by the contract and pressured by the family and goes ahead with the pregnancy despite the pain, hence, she describes the experience as "*majboori*", "I suffer so that my family is happy", "I would not have taken this big step if not for poverty". These are voices that women shared with me after several months of interaction and not the opinion one gets to hear commonly in media or on television panel discussion debates where surrogate mothers are sitting with their faces veiled along with high profile doctors, journalists, and powerful figures who instruct them before the interview on what they have to say.

By forming a community of veteran surrogate mothers, we are already giving voice to one of the more powerful voices within the surrogacy bazaar. Can we form a collective of pimps to voice suggestions whether prostitution should be legalised? In India, approximately 1.2 million sex workers are children (below the age of 18) with about 40 under-aged girls being forced into prostitution on a daily basis. With an 8% of increase in the flesh trade, India has come on the forefront as the highest trader of child prostitution. If we need to ask the most marginalized prostitutes we should ask the women/girls, who are being trafficked and are being raped and abused into forceful prostitution, who are the large majority in this practice, regarding their opinion about liberality of prostitution in India. We need to give a voice to majority of the prostitutes who are expressing themselves as being violated, and not to pimps who are party to profit-making in the business. According to Pande (2014), these so-called surrogacy "veterans" also want an increase in pay, and yearn for more contact with the intended parents, which she adds, "they very hesitantly demand". This rhetoric itself puts the surrogate mothers in a subordinate position. Why should they yearn for a continued relationship or for that matter with the children they bear, this should rightfully be their option? In my study findings, the surrogate mothers wanted a relationship with the intended parents for a continued financial and emotional support and for knowing the well-being of the child. I have hence suggested a transnational feminist alliance extending mutual support to voices against gender biased oppressions in different contexts as a way forward.

Transnational Feminism

Transnational feminism has engaged in actions across boundaries since the late nineteenth century and more recently it has been in response to globalization and capitalization that has impacted women's lives. In the 1970s and 80s women evoked a global sisterhood movement that focused on universalization which was critiqued by women of colour and women of the developing countries as Western hegemonic feminism (Weedon 1999; Grewal and Kaplan 1994; Frankenberg 1993; Mohanty 1986). This simultaneously also led to the recognition of similar hegemonies within cultural polarization based on cultural contexts (Okin 1998; Mies and Shiva 1993). The rhetoric of Western/Third World changed to global North/South, to one-third/two-third and to more neutralised social minorities/majorities (Mohanty 1986; Langlois 2009). Hence, transnational feminism has strong interconnections mainly with postcolonial feminist writings and movements. Other terminologies used are multicultural feminism which focuses on cultural determinism while transnationalism focuses on national boundaries. Since the 1980s, women have reflected on these inequalities that privilege some women over others. Recently, these transnational linkages have been based on the recognition of similarities in discriminations such as sexual service industries, women-based factories, with a commitment to common struggles (Mackie 2001). However, as can be observed, the rhetoric continues to maintain focus on colour, nation, region, and culture. Is it possible for feminism to shed some of these identities and focus on being humane? The specific forms of organizing an alliance and the praxis for feminist engagement towards it is in the process of evolving, and one suggestion would be a transnational feminist alliance with a global reproductive justice approach.

Reproductive Justice

Advocates for reproductive health, rights, and justice are increasingly aware of the safety and social risks, of ARTs and other human biotechnologies. Long-term risks of some assisted reproductive practices are under-studied, and in the US in particular, the ART field has developed almost entirely in the commercial sector and is notoriously under-regulated. Other social, ethical, and practical concerns include payments that encourage economically vulnerable women to provide eggs or to become surrogates, undertake social sex selection, and inappropriate forms of prenatal and embryo screening. Additionally, in the debate now underway on human gene editing for reproduction, the language of "choice" is sometimes misused to claim that creating a child with specified traits is the same as the right to decide whether to continue or terminate a pregnancy. Advocates of reproductive justice, health, and rights need to have a major stake in human biotechnology issues because ARTs redefine human reproduction and pregnancy and impact women's bodies, health, and well-being.

The reproductive justice framework includes women's freedom and ability to choose her sexual partner which is interfered through practices such as arranged marriage at young ages when her capacity to make such decisions is premature (Hampton 2010; Ghimire et al. 2006). It includes freedom from oppressive practices such as sati, widow cleansing, sexual assault, trafficking of women for forced prostitution, and prohibition of homosexuality. This framework aims at recognizing the histories of reproductive oppression in all communities, identifying and addressing multiple oppressions of race, class, gender, sexuality, ability, age, and immigration status to bring about changes in the structural power inequalities and develop political solidarity between women across class, race, ethnicity, sexuality, and national border. Transnational Feminism as a concept is not new and several scholars have suggested it as a way forward to build a global feminist solidarity; however, specific frameworks towards building such an alliance has not emerged within this discourse. I suggest that an alliance drawing on the reproductive justice framework that is developed by Asian Communities for Reproductive Justice (ACRJ) 2005 along with the Sister Song Collective has the potential to recognize and address inequalities and injustice observed in surrogacy practices. "Reproductive justice", hence, aims to reduce inequalities and not to use someone's vulnerability as a solution for infertility and could be a pathway towards determining global human rights and eliminating discrimination against women. However, the framework tends to weaken in the context of commercial surrogacy. The reproductive justice framework aims to assure that the genetic technologies benefit all women, their families and communities and hence falls short in giving voice to the most marginalised. Even in giving voices to the marginalized, it is usually the more powerful voices that will be heard and not the extremely oppressed ones. Taking the example of sex selective abortions most women who are extremely battered with gender bias in the society will indeed support abortions for girl child to improve their individual socio-economic status. If we justify sex selective abortions in India based on such voices that reiterate their very subordinate position as women in the society, it is futile. Can we ever ask girls who are being aborted if they are happy not to be born because the society thinks women are subordinate citizens and a socio-economic burden? Especially when, along with practices like sex selective abortions, commercial surrogacy too "joins the global roster of fungible female bodies already a favoured site of self-commodification" (Sangari 2015: 72).

Women's voices inflicting self-torture for a respect, which they regardless rightfully deserve, should not be included as representative of women's liberty or reproductive justice. Commercial surrogacy is a practice that exploits women's desperate financial needs to satisfy the reproductive desires of the elite, hence, it is a selective class-based psycho-physical violation and violence against women. The surrogate mothers in India were discriminated against by the government by unjust laws, clinics violating good medical practices, intended parents with their reproductive goals, agents for profit-making motives and also by their family members who want to make quick money using women's bodies for their benefit. I specifically question the reproductive rights of the intended parents who, in opting for surrogacy, put another woman through social stigma and psychological challenges, violate her bodily integrity and even more, put the surrogate mother's health, well-being, liberty, and even life at stake. According to Dickenson (2001), although there may be benefits of the surrogacy and increased reproductive choice for a minority few, this cannot be prioritized over the need to prevent the exploitation of some of the world's most vulnerable women. Rothman (1992) says that the woman in whom the embryo is implanted is reduced to a body part, is disrespectful to human relationships, and destructive of humanity. Let us consider the human rights pathway towards eliminating practices such as surrogacy as a discrimination against women also strongly proposed in Renate Klein's recent book on surrogacy especially considering the extent of potential harms and ethical dilemmas inherent in the practice (Klein 2017).

Surrogacy: A Human Rights Violation

Women's sexual and reproductive health is related to multiple human rights, including the right to life, the right to be free from torture, the right to health, the right to privacy, the right to education, and the prohibition of discrimination. The Committee on Economic, Social and Cultural Rights and the Committee on the Elimination of Discrimination against Women (CEDAW) have both clearly indicated that women's right to health includes their sexual and reproductive health. This means that states have obligations to respect, protect, and fulfill rights related to women's sexual and reproductive health. The Rome Petition (2017) has suggested the inclusion of surrogacy in CEDAW, by referring to the control over the reproductive rights of the surrogate mothers and all aspects of their lives during pregnancy, including threats to their physical and mental health, the primary purpose of which fulfilling the reproductive desires of another individual. This petition was developed through an international meeting in Rome, held on 23 March, 2017. Thereafter, a request was presented to the United Nations bodies, in order to address the Convention on the Elimination of All Forms of Discrimination against Women (CEDAW) and of child and human rights, to create a procedure aimed at recommending a prohibition of the practice of surrogate motherhood as incompatible with the respect of human rights and women's dignity.

Surrogacy consciously creates a state of abandonment by denying the rights of the child to know his/her origin and of the surrogate mother over the child, to know about his/her well-being and maintain a long-standing relationship. There have been children who have suffered as a result of not knowing their personal history (European Court of Human Rights 2017; http://www.echr.coe.int/Documents/FS_Childrens_ENG.pdf). The payment pattern in India is targeted towards the child and not for the reproductive services of the contract mother (Saravanan 2013). Some clinics pay the contract mothers according to the weight of the child born putting a pressure on the contract mothers to eat more. The medical practitioners fix an extra charge on the birth of every additional child (in case of twins), a small proportion of which is paid also to the contract mother as a bonus amount. Most babies born, especially of multiple pregnancies were pre-term and grossly underweight, hence, rushed immediately to the Neonatal Intensive Care Units. It is not known how many of these children survive. There are instances of disabled children born through surrogacy left in orphanages or on the streets in India. Some children were left stranded in India without any identification (passport) as their intended parents had been involved in surrogacy illegally. These children were taken care of by strangers while their parents shuttled between two countries as they could visit India only on a tourist visa (Saravanan 2013).

The Rome Petition (2017) specifies the various child rights that are violated through surrogacy. The practice violates the Convention on the Rights of the Child as per Article 7 § 1, on the child's right to know and be cared for by his or her parents; Article 9 §1, on ensuring that a child is not separated from his or her parents against their will; and Article 35, on preventing the abduction of, the sale of or traffic in children for any purpose or in any form; the Convention on the Rights of the Child on the sale of children, child prostitution, and child pornography (as per Article 2 a), which defines the sale of children as "any act or transaction whereby a child is transferred by any person or group of persons to another for remuneration or any other consideration"; Article 3, on ensuring children from sales and improperly inducing consent, as an intermediary, for the adoption of a child in violation of applicable international legal instruments on adoption are fully covered under the national criminal or penal law; Protocol to Prevent, Suppress and Punish Trafficking in Persons, Especially Women and Children, supplementing the United Nations Convention against Transnational Organized Crime (as per Article 3 a), which defines trafficking in persons as "the recruitment, transportation, transfer, harbouring or receipt of persons, by means of threat or use of force or other forms of coercion, of abduction, of fraud, of deception, of the abuse of power or of a position of vulnerability or of the giving or receiving of payments or benefits to achieve the consent of a person having control over another person, for the purpose of exploitation"; to the Convention on Protection of Children and Co-operation in Respect of Intercountry Adoption (as per Article 4, on the need to ensure that the consent of the mother, where required, has been given only after the birth of the child and that the consent has not been induced by payment or compensation of any kind, in the general spirit of the Convention); the Council of Europe Convention on the Adoption of Children (as per Article 5, on the need to ensure that a mother's consent to the adoption of her child shall be valid when it is given at such time after the birth of the child); the Council of Europe Convention on Action against Trafficking in Human Beings; the Council of Europe Convention on Human Rights and Biomedicine (Article 21); and the Petition of Fundamental Rights of the European Union, which prohibits to make "the human body and its parts as such a source of financial gain" (The Rome Petition 2017). The Rome Petition asked for a procedure to be set up for the recognition of the newborn, which shall be consistent with the rules on the rights of the child, especially with Article 7 § 1 of the Convention on the Rights of the Child, which recognizes the right of the child to know his or her mother and, as far as possible, to be cared for by her. Hence, keeping the child rights in view, surrogacy not only violates the human rights of women's bodies but also of children.

Surrogate mothers in India have expressed the separation with the child as parting with soul, hence, this procedure makes the surrogate mother's body and her very soul available to others for the production of a baby only to be given away at birth. Surrogate mothers were also concerned about the children born through surrogacy who they will never meet and get to know about for the rest of their lives. Even if this practice represents a shared complex desire for a child, it is a stratified form of reproductive labour that separates bearing and rearing of children that is differently embodied reinforcing social inequalities using people who are already discriminated historically. Women in India may express their "agency" broadly until they decide to become surrogate mothers. But within the surrogacy process, there is enough evidence that their mind, body, and behaviour is completely under the control of the clinic. In Dr. Nayna's Patel's clinic, women are held in surrogate homes and highly controlled and restricted during the entire process of surrogacy (see Chap. 4).

Their diet is controlled (they are overfed, babies are priced according to their birth-weight) and women are encouraged to read religious texts inside the clinic (SAMA and Sharma 2010). Women who are heavier are preferred for surrogacy; their consent is not taken regarding their preferred kind of relinquishment, remuneration; the child may not know the circumstances of their birth as the surrogate mothers are not allowed any involvement after birth; most babies are taken away while the surrogate mother is sedated with anesthesia; some are expected to breastfeed and be a nanny for the child(ren) and then abruptly separated; they are not insured medically; there is no life insurance; and they are paid a paltry amount in case of miscarriage. Moreover, the social context of widespread gender bias and abuse in India makes them even more vulnerable to exploitation. Needless to say, these circumstances are intrusive, deny reproductive rights and basic rights over one's own body, restricts the freedom of movement and choice to live freely as they desire (eat, read, watch television, go for a walk), hence, violates human rights. Intrusive technologies used merely to prioritize the rights of the intended parents is a violation of ethical medical practices.

Surrogate motherhood is not a technique in itself, but a practice that technology exploits within the already existing hierarchies and hegemonic systems. The technologies that were used earlier on one's own body is now used on another person's body violating their freedom dignity and integrity. For instance, in India surrogate mothers are recruited from the already vulnerable poorer communities. In the United States, although surrogate mothers are not recruited from the poorer communities in an effort to minimize risks, they are usually low-income middle class women. Even in the UK where altruistic surrogacy is allowed, there is an undeniable income disparity between intended parents and the surrogate mothers. Surrogacy goes against the UN Slavery Convention as per Article 1, which defines slavery as "the status or condition of a person over whom any or all of the powers attaching to the right of ownership are exercised", as it involves the acquisition of a right of ownership over a person and over the woman's body so as to gain possession of the child of whom she bears. There are similarities in the surrogacy practice exploiting structural inequalities throughout the world and there are possibilities to bridge the feminists thoughts in order to extend solidarity to put an end to the use of practices such as surrogacy as a option to have children.

Although the outcome of this market is justified as a win–win situation, this post-fordist industry sets up international networks primarily in countries sending intended parents for which they earn large amounts of profits and based on which surrogate mothers are recruited from countries where they are satisfied with lesser share so that middle men can make maximum profit. To take advantage of women's economic and/or social fragility in order to push them into using their reproductive capacities at the service of the rich in exchange for money is nothing but baby business. Article 3 of CEDAW states enjoyment of human rights and fundamental freedoms on a basis of equality with men and exploitation of women on the basis of their reproductive capacities is profoundly discriminatory and contrary to this objective. A decline in adoption since the technology came to use should also be a concern for all.

Many intended parents defy the law in their country or, even if it is legalized, go abroad because of lesser rights, raising questions on the humanity of such citizens who violate the rights of women in unregulated countries for their individual reproductive benefits. The Rome Petition (2017), suggests a prohibition of surrogate motherhood at the international level on the lines of female genital mutilation as a human rights violation and individuals are involved in defying the laws in the source country to adopt practices that violate human rights by misusing the inadequate regulations in the destination countries. The Rome Petition asks for a regulation to prevent people from moving from a country where surrogate motherhood is illegal to another country where it is allowed.

According to ACRJ (2005), reproductive justice is "the complete physical, mental, spiritual, political, economic, and social well-being of women and girls, and will be achieved when women and girls have the economic, social and political power and resources to make healthy decisions about our bodies, sexuality and reproduction for ourselves, our families and our communities in all areas of our lives". Reproductive oppression is "the regulation of reproduction and exploitation of women's bodies and labor as both a tool and a result of systems of oppression based on race, class, gender, sexuality, ability, age and immigration status" (ACRJ 2005: 1). Surrogacy is a practice that uses women's bodies as a tool and these women are structurally oppressed based on class, race, and gender. This definition leaves one ambiguity about what "healthy decisions" denotes. There is no mention of surrogacy in the entire report except of indirect reference to globalization which includes women working in garment and other factories, which may also relate to women risking their lives and health for bearing babies for others for a payment. Reproductive Justice framework focuses on the control and exploitation of women's bodies, sexuality, and reproduction that has been an effective strategy of controlling women and communities, particularly those of colour. Although the reproductive justice framework aims to recognize forms of oppression and to change the structural power and inequalities as well as defines some strategies, this definition cannot be applied to the complex forms of inequalities in surrogacy practices. The strategies suggested in ACRJ 2005 report on reproductive justice is focused on immigration based policies in Canada. Hence, reproductive justice strategies in the context of surrogacy can only be derived from this report. The Rome Petition comes the closest to drawing strategy pathways towards global assisted conception most relevant to reproductive justice. The next chapter, attempts a way forward towards a humanitarian assisted conception that respects human dignity and integrity.

I conclude this chapter with two humanitarian questions that I will continue to discuss in the next chapter:

1. When we consider altering of genomes as undesirable and violating human worth, how can we, as humans even consider the normalization of picking gametes, choosing surrogate mothers and paying for babies based on classist, racist, ableist, and other social prejudices?
2. How can we normalize surrogacy, a practice that is an intrusion to another woman's reproductive being and that can risk her health, well-being, and life as "any other work"?

References

ACRJ. 2005. *A new vision for advancing our movement for reproductive health, reproductive rights and reproductive justice*. Oakland, CA: Asian Communities for Reproductive Justice.

Bailey, Alison. 2011. Reconceiving surrogacy: Toward a reproductive justice account of Indian surrogacy. *Hypatia* 26: 715–741.

Callahan, Joan C., and Dorothy E. Roberts. 1996. *A feminist social justice approach to reproduction-assisting technologies: A case study on the limits of liberal theory*. Faculty Scholarship. Paper 1155.

Cherry, April L. 2014. The rise of the reproductive brothel in the global economy: Some thoughts on reproductive tourism, autonomy, and justice. *17 U. Pa. J.L. & Soc. Change* 257: 269–271.

Dickenson, Donna. 2001. Property and womens alienation from their own reproductive labour. *Bioethics* 15 (3): 205–217.

Donchin, A. 2010. Reproductive tourism and the quest for global gender justice. *Bioethics* 24 (7): 323–332.

European Court of Human Rights. 2017. *Strasbourg*. http://www.echr.coe.int/Pages/home.aspx?p=home.

Fixmer-Oraiz, Natalie. 2013. Speaking of solidarity: Transnational gestational surrogacy and the rhetorics of reproductive (in)justice. *Frontiers: A Journal of Women Studies* 34 (3): 126–163.

Frankenberg, Ruth. 1993. *White women, race matters: The social construction of whiteness*, 304. University of Minnesota Press.

Gaard, Greta. 2010. Reproductive technology, or reproductive justice?: An ecofeminist, environmental justice perspective on the rhetoric of choice. *Ethics & the Environment* 15 (2), 103–129.

Galpern, Emily, Mia Mingus, and Cara Page. 2007. Surviving ourselves: Unpacking reproductive justice, genetic technologies and eugenics. In *Presentation for SisterSong's "Let's Talk About Sex" Conference*, Chicago, IL, June 1, 2007.

Ghimire, D.J., W.G. Axinn, S.T. Yabiku and A. Thoirnton. 2006. Social change, premarital nonfamily experience, and spouse choice in an arranged marriage society. *American Journal of Sociology* 111: 1181–1218.

Grewal, I., and C. Kaplan. 1994. *Scattered hegemonies: Post-modernity and transnational Feminists Practices*. Minneapolis: University of Minnesota Press.

Gupta, J. 2012. Reproductive biocrossings: Indian egg donors and surrogates in the globalized fertility market. *IJFAB: International Journal of Feminist Approaches to Bioethics* 5: 25–51.

Hampton, Tracy. 2010. Child marriage threatens girl's health. *JAMA, The Journal of the American Medical Association* 304 (5): 509–510.

Jesudason, Sujatha, and Katrina Kimport. 2013. Decentering the individual and centering community: Using a reproductive justice methodology to examine the uses of reprogenetics. *Frontiers: A Journal of Women Studies* 34 (3): 213–225.

Klein, Renate. 2017. *Surrogacy: A human rights violation*. Melbourne: Spinifex Press.

Langlois, A. 2009. Normative and theoretical foundations of human rights. In: Human rights: Politics and practise, ed. by M. Goodhart. Oxford: Oxford University Press.

Luna, Zakiya, and Kristin Luker. 2013. Reproductive justice. *Annual Review of Law and Social Science* 9: 327–352.

Mackie, Vera. 2001. The language of globalization, transnationality and feminism. *International Feminist Journal of Politics* 3: 180–206.

Mies, M., and V. Shiva. 1993. *Ecofeminism*. London: Zed.

Mohanty, C.T. 1986. Under western eyes: Feminist scholarship and colonial discourses. *Boundary* 2 (12): 333–358.

Mohanty, C.T. 2003. "Under western eyes" revisited: Feminist solidarity through anticapitalist struggles. *Signs* 28: 499–535.

Mohapatra, Seema. 2012. Achieving reproductive justice in the international surrogacy market. *Annals of Health Law* 21 (1): 191–200.

Okin, Susan Moller. 1998. Feminism, women's human rights, and cultural differences. *Hypatia* 13: 32–52.

Pande, A. 2014. *Wombs in labor: Transnational commercial surrogacy in India*, 252 p. New York: Columbia University Press.

Roberts, Dorothy E. 2009. Race, gender, and genetic technologies: A new reproductive dystopia? *Signs* 34 (4): 783–804.

Rothman, Barbara Katz. 1992. Reproductive technologies and surrogacy: A feminist perspective. *Creighton Law Review* 25: 1599–1616.

Rudrappa, S. 2015. *Discounted life: The price of global surrogacy*. New York: New York University Press.

SAMA and S. Sharma. 2010. *Can we see the baby bump*. India: Surfilms.

SAMA. 2012. *Birthing a market: A study on commercial surrogacy*. New Delhi: SAMA-Resource Group for Women and Health.

Sangari, Kumkum. 2015. *Solid liquid: A (trans)national reproductive formation*. Tulika Books: New Delhi.

Saravanan, S. 2010. Transnational surrogacy and objectification of gestational mothers. *Economic and Political Weekly* 45: 26–29.

Saravanan, S. 2013. An ethnomethodological approach to examine exploitation in the context of capacity, trust and experience of commercial surrogacy in India. *Philosophy, Ethics and Humanities in Medicine* 8: 10.

Scheper-Hughes, N. 2000. The global traffic in human organs. *Current Anthropology* 41 (2): 191–224.

Sunder Rajan, Rajeswari. 2000. Introduction: Feminism and the politics of resistance. *Indian Journal of Gender Studies* 7: 153–165.

The Rome Petition. 2017. 2nd NGO European meeting on the abolition of surrogacy. 27th February. European Women's Lobby: piazza di Monte Citorio, Rome.

Vora, Kalindi. 2009. Indian transnational surrogacy and the commodification of vital energy. In special issue, Re-tooling subjectivities: Exploring the possible through feminist science studies. *Subjectivities* 28 (1).

Weedon, Chris. 1999. *Feminism, theory and the politics of difference*. Oxford: Blackwell.

West, Robin. 2009. From choice to reproductive justice: De-constitutionalizing abortion rights. *The Yale Law Journal* 118: 1394–1432.

Chapter 8
Towards Humanitarian Assisted Conception

Abstract Roberts (Signs 34:783–804, 2009) notes reproductive liberty must encompass not merely the individual women's choice about how and when to have a child and to end her pregnancy, but should also include social justice. The notion of liberty allows an individual to choose procreation for oneself but without causing any harm to others and also includes freedom of choice in abortion (Bolton in United States Reports 179, 1973). The same logic when applied to surrogacy involves an inherent tampering of another person's reproductive right to abortion and parenthood impacting their health and well-being. Yet when it comes to surrogacy, liberals have demanded for evidence that the practice devalues individual reproductive rights. Several scholars have noted that individual reproductive rights come along with responsibilities towards a just and humane society. One way forward is to examine the humanitarian reproductive thresholds based on social justice. Drawing on intersectionality, the aim in examining humanitarian thresholds is to identify the humane responsibilities and threshold that may be crossed in asserting individual reproductive desire for children through surrogacy. I will discuss the two questions that I raised in the previous chapter; as humans can we normalize gamete picking and surrogate choosing that reify social prejudices, and can we normalize surrogacy as any other work? I begin with defining humanitarian threshold.

Keywords Human rights · Intersectionality · Geneticization · Surrogacy as work Humanitarian conception

I define humanitarian reproductive thresholds as reproductive rights that is applicable to one's own body and its limitations when it is applied on another person's body. I should have complete control over my own reproductive body but that should not include a rightful control over another person's reproductive functions and rights. Any form of individual reproductive desires that interferes with another person's reproductive labour such as genetic substance retrieval, bearing, or rearing cannot be defined as a reproductive right and crosses the humanitarian thresholds.

S. Saravanan, *A Transnational Feminist View of Surrogacy Biomarkets in India*,
https://doi.org/10.1007/978-981-10-6869-0_8

The aim in examining humanitarian thresholds is to identify the humane responsibilities that may be crossed in asserting reproductive desires. Individual reproductive rights come along with responsibilities towards a just and humane society. Petchesky (1995) reiterates the misconception of women owning their own body that turns it into a reproductive factory by objectifying and denying it basic dignity: "the individual right of woman over her own body" as "an unconscious mirroring of the capitalist-patriarchal ideology … premised on the logic of bourgeois individualism and inner urge of private property" (Farida Akhter, cited in Petchesky 1995).

Drawing on this individualised misconception of woman's own body, I aim to explain what I mean by humanitarian thresholds and, in the process, attempt to determine what reproductive rights do individuals not have on other people's bodies. Infertility is known to put people through social stigma, psychological problems, physical stress of infertility treatment, and violation of bodily integrity. But by opting for surrogacy in order to fulfil the desire to have a child, the intended parents are inclined to put another woman through the same set of problems; social stigma, psychological challenges, violation of her bodily integrity, and even more, putting the surrogate mother's health, freedom, liberty and even life at stake. Hence, the surrogacy arrangement clearly crosses the "humanitarian" threshold of the very ideologies that feminism and reproductive justice itself stands for which is "procreation for oneself but without causing any harm to others".

Surrogacy and Intersectionality

As a violation of basic human rights and from a child right's perspective, I contend in line with the Rome Charter, of which I am also a signatory, that surrogacy should be globally prohibited. For any individual experimenting with IVF procedures on their own body is a matter of choice although it is invasive and harmful to a woman's body and self; these rights need to be restricted to the person carrying out reproductive interventions on one's own body. When this reproductive right extends to a control over someone else's body it is an intrusion into another person's body to have children. When "I" (intended parent) can control "your" (surrogate mother's) body, womb, mind, possible emotions towards the child and life and the resulting child, then "my body and my own" logic stands void. Surrogacy practice involves a control of another human being's body to fulfil one's own reproductive desires and this procedure violates another person's human dignity and reproductive rights. The reproductive labour of bearing a child itself is supposed to give an individual (generally a woman) certain basic human rights but the surrogacy process inherently restricts and violates these particular rights of the surrogate mother. It restricts the abortion rights of another woman, it restricts another person's liberty to experience parenthood to its full extent, it restricts the woman's right over the child, and moreover it puts another person through the physical and psychological pains of the IVF procedure, bearing a child, the birth and after-birth. The surrogacy process violates these rights, and as mentioned earlier it is likely to put another woman (the surrogate mother) through social stigma, psychological challenges, violation of her bodily integrity, and also put the surrogate mother's health, freedom, liberty and even life at stake. Hence, surrogacy cannot be considered a socially justified practice and it crosses the threshold of the humanitarian reproductive rights.

Intersectionality includes other complexities inherent in the surrogacy process. Co-parenting and liberal reproductive rights, as specified by Teman can maintain human dignity only if such contracts take place between two equals (Teman 2010). But it is known that surrogacy contracts between equals are almost non-existent in surrogacy, even in cases of altruistic surrogacy. Intersectionality is a structural approach that examines simultaneous intersections of social difference and identity (such as those related to meanings of race/ethnicity, gender, class, sexuality, age, disability/ability, migration status, and religion). This should also include forms of systematic oppression, colonialism and also caste in India. The models of intersectionality proposed from within feminist scholarship elucidate that the gendering of reproductive substance and labour, is closely entwined with race, nationality, or class. This scholarship focuses on the experiences of the most marginalized populations in comparison and contrast with those who are most privileged. As intersectionality theory reminds us, human beings experience different degrees of penalty or privilege depending on their social location (Dhamoon and Hankivsky 2011). It informs the multiple ways by which systems of oppression combine to produce marginalizing effects. It refers to forms of relationship between social structures that combine to create social categories to which certain experiences and forms of expression are unique (Cho et al. 2013). Speaking about IVF technologies, it may at first glance seem to promote gender equality by enhancing women's choices. But by introducing class to the analysis, a different pattern of differential choices emerge. Further, an inclusion of race, ethnicity, and migration status reveals various forms of domination and discrimination.

CEDAW, hence, needs to include right to abortion, right to parenthood for all women regardless of whether it is a genetic or gestation link. Racism, classism, ableism, and patriarchy has to be ended through open resistance and not through hidden laboratory practices offering anonymous donors or parenthood including surrogate mothers alongside a laissez-faire gamete choice, sex selection, embryo choice, pre-implantation genetic diagnosis (PGD), and finally, technologies that will tamper with genetic components of the embryos. This is a one-sided laissez faire market catering to the intended parents and indirectly assisting a reinforcement of all the unjustified social prejudices that feminism has been striving to resist for all these decades. Science and technology has brought about wonderful changes in every walk of life. But most of the ARTs are based on dishonesty, lies, and secrecy. Anonymous gamete is sought so that parents don't have to inform their forthcoming child the truth about their birth. Affluent people are hopping from one place to another seeking surrogacy in unregulated countries with high incidence of poverty to ensure easy exit with their babies and to find women unprotected by law and willing to do anything for money.

Surrogacy should be disallowed in any circumstances because even if a woman is not donating part of her body or her biological substance, she is creating another human being and then giving it away. It is not merely about the surrogate mother who takes a decision about her own body but she is bringing another individual into this world and is signing away her rights towards that child whose opinion has not been taken into account. Moreover, it's not enough to blame it on overstimulation or maternal mortality that may cause a woman's death during surrogacy or egg donation as this process has been initiated to fulfil the desires of someone else's reproductive desire, this practice that causes life risk should not be allowed on humanitarian grounds.

Individual parental desires and procreative choices are determined by the symbolic body; social position, group membership, social location, and life prospects. These desires are also determined by the body self-mind experiences. However, medical practitioners in India tend to relate it with skin-deep factors such as skin colours by saying that surrogate mothers do not and cannot bond with a child of different skin colour. A Marathi film in 2011 "*Mala Aai Vhhaychay*" meaning "I want to be a mother", is a story about a surrogate mother who enters into a surrogacy contract to enhance her household economic condition (Porey 2011). She refuses to abort the child after it is found that the child could be born with disabilities. The intended mother abandons the baby and leaves India. This story about an Indian mother bonding with a white skinned child is inspired by a true story and reveals the complexity of surrogate motherhood. A feminist social justice approach understands the harms of reproductive-assisted technologies caused by contextualized and socially constructed societal positions (infertility, parenthood, motherhood) (Callahan and Roberts 1996). The harm caused by certain technologies to the symbolic body is exemplified in subordinate groups and unjust social structures.

At a point of time when feminists support practices such as sex selective abortions and female genital mutilation in Asia and Africa, such thoughts processes are already becoming inhuman. It is almost equivalent to saying that violence against women and child abuse is acceptable because it is socially sanctioned in some parts of the world. When researchers say that selective abortion in Asia is sexist while in America it is family balancing, thoughts are getting dangerously close to racist hypocrisy. Female genital mutilation is banned in most of the developed countries, but when the developing countries try to prohibit this practice it is seen as an imposition of external influence. Similarly, when feminists argue that it is justifiable for a poor woman desperately in need for money to carry a baby for another (wo)man and that through the process they are emancipated by not bonding with the child, such thoughts are inhuman. Most developed countries prohibit surrogacy, but when developing countries do the same, there is a cry that the respective government is depriving the constitutional rights of individual surrogate mothers to earn some extra money to bring their families out of poverty. Countries that prohibit commercial surrogacy within their borders still allow people to go and exploit poor citizens of others countries and bring back the children into their borders. Post-earthquake in Nepal, the Israeli government sent a plane to bring back the Israeli babies and left behind surrogate mothers in the disaster zone.

One would think that feminists would not subscribe to these thoughts, but in recent times, some of the feminist discourse reflect these insensitivities towards fellow human beings in general and women in specific. What follows in this chapter is an attempt to identify "humanitarian" thresholds of these feminist ideologies using a case of surrogacy arrangement in India and its global relevance to reproductive justice. All kinds of **phrases** are used, also by academics to separate the surrogate mother from the baby in her womb, "surrogate", "womb mothers", and her reproductive organ is called a "vacuum", "a container". In the minds of those using this bio-market, these women are machines until the baby is born; if there were artificial wombs available in this world, human bodies would not have carried the babies. So do the women feel like machines; maybe some do when they are carrying the baby, but when the baby is born they are real human beings and many women bond with the babies almost instantly. This is clear in the reaction of the women; some who know that they would never see the baby again put their hand on the baby's head to bless the child with tears flowing down their eyes, those who care for the babies on birth are kissing cuddling, breastfeeding the babies only to face separation with no further contact, while yet others who never get to see the face of the child are left with a life-long urge of seeing the babies they have borne. A surrogate mother told me that her only wish was that the photos of her twins be sent to her every year so that she could see how they are growing. Is this similar to a person working in a bag factory longing to have a look at the bag(s)he has produced years after she produced it and whether the person who owns it is caring for this bag or not? Is this entire practice inhuman or are we becoming selectively mechanical in this world of technological developments wherein some lives, emotions, and opinions are highly valued while others are trivialized? Researchers keep track of their publications, every piece we write is our baby, an idea we conceived, we want to receive feedback, and see our concepts grow or being critiqued. What if we are not allowed any access to our research soon after we publish and we have to sign such a contract with our publishers?

At a recent conference I attended which was meant to be an academic gathering to discuss all viewpoints, I realized most of the speakers were supporting surrogacy. After my presentation, one of the question I was asked was "it looks like you have interviewed only one woman because what you are presenting seems very dramatic?". I was asked another question regarding the eating habits of the surrogate mothers in surrogate homes in India and when I began answering about how the final payment is made based on the weight of the baby, I was stopped mid-way saying my time was up. In this workshop environment in Paris where everything was discussed in terms of individualization, I seemed to be the only one stirring up the calm waters by talking about maternal emotions, poverty, and exploitation. The participants in the gathering were unhappy with me for bringing up the dirt that had settled below. Similarly, women who experience near-death situations or even die of egg donation or the surrogacy procedure anywhere around the world are given the least attention by the media. Every other household in India is employing child labour to help with household chores, every other restaurant has children working overtime without a chance to balance school and work. When children can be used in the markets, why not adult women who are willing participants offering themselves to be scapegoats? Yet, Dr. Nayna Patelsays "*As soon as you come to India, you talk about exploitation*" (The VJMovement 2009). The other argument supporting surrogacy is that surrogacy is just like any other work (Pande 2014; Rudrappa 2015). I argue that surrogacy is neither a technology nor is it a work, it is a practice that intrudes into women's body through technology embedded in class and gender inequalities using historic means of exploitation.

Surrogacy as Work

In this normalization of exploitation, one of the arguments supporting surrogacy is that people should have the right to choose their own employment, equating surrogacy with works such as those involved in garment factories, prostitution, or are housemaids (Pande 2014; Rudrappa 2015). Authors who presume this have compared surrogacy in India with garment factory workers in justifying that surrogacy work is better than garment factories. The same scholars also agree that the working conditions in garment factories is an outright human rights violation of labour. By justifying surrogacy work as a better option than garment work, the scholars are normalizing the working conditions of garment factories and adding surrogacy to this work.

Regardless of whether it is neocolonialism or postcolonialism, such labour categorized as informal employment are exploitative and moreover are specifically women-based labour. Of these, prostitution is sexually intrusive and surrogacy is an intrusion into women's reproductive self. These are practices that are strengthening the control of the poor by the rich. Focus should be diverted towards offering enhanced working conditions for garment workers and housemaids and such forms of women-based labours so that women do not have to sell their reproductive capacity. The argument of surrogacy as work emphasizes that women are paid for their reproductive labour. My studies and several other studies have found that the payment is made mainly for the child, precisely for handing over the child. Some scholars claim that the surrogate homes are good for surrogate mothers providing agency and friendships (Pande 2014; Rudrappa 2015). These assumptions are also adopted by others saying that women are provided with healthy food, medical support, computer classes, English lessons and other forms of vocational classes (Sifris 2015). When I visited the same clinic, there was no sign of any such classes and I was told by the doctor that the surrogate mothers were not interested so they stopped it. It's a lame excuse as the surrogate mothers inside the homes are constantly in rotation, in and out and so are their needs. Anyways, this is beside the point and offers no justification for women to be detained without their wishes in dormitories. It is rather the thought process that these women eat unhealthy food and have dangerous bodies that need to be monitored throughout their pregnancy that is problematic. I argue that mandatory detainment of women and the kind of rules imposed within these dormitories is a violation of human rights and should not be allowed under any circumstances. Women who were begging to go home were not allowed to do so in my study. Surrogacy is an alienated form of labour and contextually women in India made reference to giving away the child as relinquishing a part of their soul, which stands in contrast to what is normally reported in media or by many scholars in India. The surrogate mothers experience this embodiment despite being constantly reminded by the clinic personnel that the child belongs to the intended parents and they are only renting their wombs, and that they are being paid for their service like any other work. The intended parents accept at face value whatever is told to them by the medical practitioners as being the reality and claim that the child belongs to them because the pregnancy was their intent, as they had purchased/provided the gamete and paid for the pregnancy.

It is argued that surrogacy is a "win-win situation", which is Dr. Nayna Patel's popular dialogue also adopted by several academics (Rowley 2013). If I review this win-win argument with my research findings, I find that the intended parents may experience intrusive technologies and receive a desired child by paying a monetary price, while the surrogate mothers receive monetary returns that helps their family in a big way in return for an extreme form of bodily intrusion, alienation from their own children and family members, psychological pain, and for most of them, having to forego a part of their soul. Some surrogate mothers gain respect from their husband and extended family while others face familial and social ostracizing because of their involvement in surrogacy. This respect she could have earned if she had been treated with equality and had been educated enough to earn a respectful income in the first place. My study reveals that while women may be respected, they are primarily treated as a child-making machines as surrogate mothers. Most women are expected to repeat surrogacy to fulfil the familial financial target and this may be a fixed amount needed to bring the family out of poverty or education, to fulfil monetary needs for health and dowry purposes, to buy a house, land or clear previous debt and, in the process, some surrogate mothers experience near death situations and some have to forgo their uterus or experience serious morbidities given their substandard reproductive health to begin with. Finally, I take the genetic normalization argument of reproductive technologies that also cross humanitarian thresholds.

Genetic Normalization Through Surrogacy Practice

Human rights help to ensure respect for individual worth and the common good in the face of powerful biotechnologies. Hence, claims to universal human rights depend also on the formal recognition of humanity. Many countries and international declarations have used human rights as a framework to establish policies regarding human biotechnologies. For example, the Council of Europe's Convention on Human Rights and Biomedicine and UNESCO's Universal Declaration on the Human Genome and Human Rights, reject biotechnology applications that alter the genomes of future generations. This brings me to the other question I raised at the end of my previous chapter; when we object to genetically modified human beings, can we then normalize the picking of gametes, choosing surrogate mothers and paying for babies based on classist, racist, ableist characteristics that reinforce social prejudices. Genetic engineering has already arrived with scientists in the US reportedly having edited the genes of human embryos using CRISPR (clustered regularly interspaced short palindromic repeats) (NDTV 2017). The article observes that this technology may open gates to designer babies, but we have been supporting technologies that subtly lead towards the designer baby concept. The liberals may argue when we can pick and choose gametes, why not modify the genes to suit the individual benefit and desires of the intended parents. One intended parent said in the documentary "*Future Baby*", "As genetic engineering is available, parents would want to throw-in a trait or two" which sounded more like ingredients in a cooking recipe rather than a child. He added, "When I can have a perfectly wonderful healthy baby, why gamble?" (Arlamovsky 2016). In this age of rapidly advancing scientific revolution, we as human beings need to take a step back and revisit our past mistakes and think what sort of world we are willing to leave behind as legacy to the future generations. If we gauge our future children with their race and abilities that emphasize on the superiority of some races and abilities over others, we are leaving a legacy that will find logic in waging war against each other to maintain the superiority of one section of people over others.

Throughout this book, we have seen examples of gamete picking that are individual choices that may seem liberal but have some irrevocable social effects. By genetic normalizing, I mean giving precedence to genetic component such as phenotype and ability. Those using IVF treatment very casually come across a range of market choices of gamete donors and surrogate mother. One intended parent in USA said, "our donor is a Brazilian model" (Arlamovsky 2016). We have gradually begun to normalize and internalize such comments and dont see anything strange in it anymore. The egg donor of the same couple says, "I would like to erase the thought that my egg donation is resulted in a child, I am worried what if the child wants to know the biological mother. I am doing this only for economic reasons". Phenotype choices feed into racist kinship sentiments. "In the new dystopia, genetic selection technologies that incorporate race as a biological category reinforce class divisions between elite people of colour who can afford the full array of high-tech procedures and the masses who suffer most from neoliberal policies bolstered by these very biological explanations of racial inequities" (Roberts 2009).

While in surrogacy, the genetic links are prioritised over gestational role, in anonymous gamete donation, it is the buyer that is given precedence over the seller. Several egg donors have expressed guilt or concerns regarding the child that would be born through their donation. Liberals tend to support anonymity in the process of surrogacy and gamete donation prioritising the reproductive rights of the intended parents. But in practical terms, there are several other human beings and parents involved in such a process, a gamete donor is still called the biological parent, the surrogate mother is the birth mother who plays a significant role in the birth of the child for the intended parent, hence, their role cannot be anonymized both from the parental or child's point of view. On the question of anonymity, one doctor in a documentary film opined that we are entering a phase where information about parents in not needed, two men and two women are parenting children. He questioned, "Does a child need a father and a mother, or even parents?" (Arlamovsky 2016). Another girl child now grown-up in Israel in the same documentary born to a single mother through donor insemination disappointedly expressed, "The most difficult part is being unable to know my father. It's unfair that I will never know my biological father or my half brothers and sisters. I can't meet my half-brothers, even if we wanted to we couldn't. All I know about my father is his hair colour and height, no name no picture". She concluded by saying, "I think that's selfish, I think there should be a choice open for children", and rightly so (Arlamovsky 2016). If we can recall in Chaps. 5 and 6, almost all the surrogate mothers I spoke to in India also had the same concern, they were worried that their children will have half-siblings somewhere in the world whom they will never be able to meet. The mother of this child said that "there is a lot of ignorance about all this as people don't understand (the complexity) and a lot of mothers don't want non-anonymity because they are scared that this child would go away" (Arlamovsky 2016). The mother who was confused at this situation said, 'I thought if I love the child enough that would be enough'. From a child right's perspective, I contend that anonymous gamete donation should be disallowed. The two reasons why intended parents want anonymous donors is because they want to keep the donor a secret from their child because they are scared of losing the child and hide their infertility from the society. Both these motives do not prioritise the child's interest.

Moreover, liberating and normalizing this technology involves processes such as picking and choosing of gametes, reifies the very barriers of race and ethnicity that we are meant to erase. Selection criteria on gametes initiates a chain of social injustice through gamete markets and reinforces societal prejudices. I do not mean to say that an individual sitting and selecting a gamete donor is inherently a racist but by undertaking such a selection process this person is automatically initiating a whole range of market process behind the computer. One click of selection on the computer based on desired criteria causes market changes. It is affecting the selection criteria of donors making some donors more bioavailable based on their skin colour, ability, and many other phenotypes that may be tagged with a higher biovalue over others. Some may argue that it is acceptable for individuals to select gametes based on their own phenotypes as it is socially acceptable for the child and for themselves that they raise children looking like them. I ask, who determines this social acceptability; if more and more people with a particular skin colour had children with mixed skin colour, it would become a norm and then the social acceptability would be transformed. Human contribution should not be determined by their genetic component, skin colour, phenotype, or ability. The scarcity of gametes is a pretense, there is an excess of biomaterial stored in laboratories around the globe. In the documentary film "*Future Baby*", one company storing gametes said they have stored gametes for years because they are not allowed to dispose any and they know most of the gametes they have stored is of people who have already died (Arlamovsky 2016).

Robertson centres his debate on the human desire to take their generation forward or have a genetic connection with a child. This is a social normalization of geneticization process that values people to their ability to carry forward their filiality, devalues the gestational link between the surrogate mother and child, and devalues all the children who are not genetically connected with their parents and their parent–child relationship. Defining parenthood genetically devalues several other human beings, hence causing inherent harm to other fellow humans. I may be considered illogical in suggesting "humanitarian assisted conception" as a way forward, and with socially prejudiced individualistic choices increasingly becoming normalized, my suggestion may sound abstract. However, such individualist choices may seem as if we are modernizing, but actually we are going backwards. We sympathize with homosexual rights and in the same breath normalize the exploitation of poor women's bodies on account of their vulnerability to bear children for the rich. Feminists have supported sex selective abortions and female genital mutilations with cultural determinism. Such sympathy is clearly selective along well-established global hierarchies reinforcing the rich, white, patriarch, and ableistic superiority. We are heading towards a reality that is not far from Margaret Atwoods' fiction *The Handmaid's Tale*. Seemingly liberal at the first instance but we are in fact going backwards.

References

Arlamovsky, Maria. 2016. *Future baby*. NGF: Nikolaus Geyrhalter Filmproduktion.

Bolton, Doe V. 1973.410. United States Reports 179.

Callahan, Joan C., and Dorothy E. Roberts. 1996. A feminist social justice approach to reproduction-assisting technologies: A case study on the limits of liberal theory. *Faculty Scholarship*. Paper 1155

Cho, Sumi, Kimberlé Williams Crenshaw, and Leslie McCall. 2013. Toward a field of intersectionality studies: Theory, applications, and praxis. *Signs* 38 (4): 785–810.

Dhamoon, R.K., and O. Hankivsky. 2011. Why the theory and practice of intersectionality matter to health research and policy. In *Health inequities in Canada: Intersectional frameworks and practices*, ed. O. Hankivsky, 16e50. Vancouver: University of British Columbia Press.

NDTV. 2017. First genetically modified human embryos in the US may open gates to 'designer babies'. 27 July, New Delhi Television: New Delhi.

Pande, A. 2014. *Wombs in labor: Transnational commercial surrogacy in India*. New York: Columbia University Press.

Petchesky, R. 1995. The body as property: A feminist re-vision. In *Conceiving the new world order: The global politics of reproduction*, ed. F. Ginsburg and R. Rapp. Los Angeles, CA: University of California Press.

Porey, Samruoddhi. 2011. Mala AaiVhhaychay.

Roberts, Dorothy E. 2009. Race, gender, and genetic technologies: A new reproductive dystopia? *Signs* 34 (4): 783–804.

Rowley, Tom. 2013. The British babies made in India. *Telegraph.*

Rudrappa, S. 2015. *Discounted life: The price of global surrogacy.* New York: New York University Press.

Sifris, Ronli. 2015. Commercial surrogacy and the human right to autonomy. *Journal of law and medicine* 23: 365–377.

Teman, Elly. 2010. *Birthing a mother: The surrogate body and the pregnant self.* Berkeley: University of California Press.

The VJMovement. 2009. Rent-a-womb: Outsourcing surrogacy in India. *YouTube.* https://www.youtube.com/watch?v=GBHs6OI6xIs&t=1s. Accessed 14 Aug 2017.

The manufacturer's authorised representative in the EU is Springer Nature Customer Service Centre GmbH, Europaplatz 3, 69115 Heidelberg, Germany. If you have any concerns regarding our products, please contact ProductSafety@springernature.com

Printed and bound by CPI Group (UK) Ltd, Croydon, CR0 4YY
25/09/2026
02222898-0003